SpringerBriefs in Applied Sciences and Technology

SpringerBriefs present concise summaries of cutting-edge research and practical applications across a wide spectrum of fields. Featuring compact volumes of 50 to 125 pages, the series covers a range of content from professional to academic.

Typical publications can be:

- A timely report of state-of-the art methods
- An introduction to or a manual for the application of mathematical or computer techniques
- A bridge between new research results, as published in journal articles
- A snapshot of a hot or emerging topic
- An in-depth case study
- A presentation of core concepts that students must understand in order to make independent contributions

SpringerBriefs are characterized by fast, global electronic dissemination, standard publishing contracts, standardized manuscript preparation and formatting guidelines, and expedited production schedules.

On the one hand, **SpringerBriefs in Applied Sciences and Technology** are devoted to the publication of fundamentals and applications within the different classical engineering disciplines as well as in interdisciplinary fields that recently emerged between these areas. On the other hand, as the boundary separating fundamental research and applied technology is more and more dissolving, this series is particularly open to trans-disciplinary topics between fundamental science and engineering.

Indexed by EI-Compendex, SCOPUS and Springerlink.

Surachate Kalasin

Nanoscale Lab-on-a-Chip Sensors

Healthcare Applications

Springer

Surachate Kalasin ⓘ
Nanoscience and Nanotechnology
Graduate Program
Faculty of Science
King Mongkut's University of Technology
Thonburi
Bangkok, Thailand

ISSN 2191-530X   ISSN 2191-5318   (electronic)
SpringerBriefs in Applied Sciences and Technology
ISBN 978-981-96-5980-7   ISBN 978-981-96-5981-4   (eBook)
https://doi.org/10.1007/978-981-96-5981-4

© The Editor(s) (if applicable) and The Author(s), under exclusive license to Springer Nature Singapore Pte Ltd. 2025

This work is subject to copyright. All rights are solely and exclusively licensed by the Publisher, whether the whole or part of the material is concerned, specifically the rights of translation, reprinting, reuse of illustrations, recitation, broadcasting, reproduction on microfilms or in any other physical way, and transmission or information storage and retrieval, electronic adaptation, computer software, or by similar or dissimilar methodology now known or hereafter developed.
The use of general descriptive names, registered names, trademarks, service marks, etc. in this publication does not imply, even in the absence of a specific statement, that such names are exempt from the relevant protective laws and regulations and therefore free for general use.
The publisher, the authors and the editors are safe to assume that the advice and information in this book are believed to be true and accurate at the date of publication. Neither the publisher nor the authors or the editors give a warranty, expressed or implied, with respect to the material contained herein or for any errors or omissions that may have been made. The publisher remains neutral with regard to jurisdictional claims in published maps and institutional affiliations.

This Springer imprint is published by the registered company Springer Nature Singapore Pte Ltd.
The registered company address is: 152 Beach Road, #21-01/04 Gateway East, Singapore 189721, Singapore

If disposing of this product, please recycle the paper.

To mom, my wife Ann and my child Nita without whom this textbook would never have been completed.

In loving memories of dad, younger brother, and bichon Boonggie, whose stories live on pages, until our paths cross once again.

Foreword

It is undoubtedly a very challenging and dedicated work to write a textbook on fundamental nanoscience as well as advancement of nanotechnology that incorporate the most recent developments in lab-on-a-chip technologies and wearable sensors for smart healthcare, including recent articles, as it integrates several sophisticated multidisciplinary domains to compile. The developed structures presented here are all categorized based on features that are particularly relevant to applications in biotechnology, biosensing, electrochemistry, molecular simulation for sensing applications, biomedicine, diagnostics, analytical biochemistry, polymers for nanotechnology, self-sensing intelligent microrobots, and wearable sensor development for telemedicine healthcare.

The ideas that originate this textbook came up after the publication of "Challenges of Emerging Wearable Sensors for Remote Monitoring Toward Telemedicine Healthcare" (*Anal. Chem. (2023), 95, 3, 1773–1784*). Even there was the emergence of the Internet of Things (IoT), smart healthcare remains disconnected from the global Internet technology because health monitoring devices are often expensive, large, and reserved for usage by medical professionals. These exciting wearable sensors could decentralize health monitoring from a hospital setting to a point-of-care (POC) diagnostic made at home. The author also foresees that the development of non-invasive lab-on-a-chip technologies may lessen multimorbidity, which is a risk factor in today's aging population. Looking back over the past decade, wearable technology trends that involve significant increases in both research and commercialization may be observed. Nevertheless, advancements and setbacks have contributed to the success of wearable sensors and lab-on-a-chip technology. Most commercial advancements in point-of-care health monitoring rely on clever adaptations of current mechanical, electrical, and optical techniques. The innovation in downsizing biocompatible sensor devices for human bodies is another aspect of the adaption. However, creating multiplexed wearable sensors for telemedicine and healthcare are also tremendous challenges as developing chemical sensing multimodalities. Besides, telehealth wearable sensor with the satellite communication for point-of-care monitoring in real-time is also explained in another article (*ACS Biomaterials Science and Engineering, (2020), 7, 1, 322–334*).

As the aging population increased, long-term care for multimorbidity has been recognized as a global concern with substantial effects on individuals, caregivers, and society. Multimorbidity has been related to poor health conditions and a high risk of mortality that can result in an overload of contemporary healthcare management. Noninvasive multifunctional wearable technology with long-term care provides a promising future for personalized and remote health assessment that enables delay or mitigate epidemiology of multimorbidity, which is susceptible in aging societies. The multiplex wearable sensor to monitor heart and kidney condition using sweat is also delved into my article (*ACS Applied Nano Materials, (2023), 6, 19, 18209–18221*). With the advancement of nanotechnology, lab-on-a-chip innovation can be transformed as an artificial intelligence-based workforce as declining birth rates and aging populations will likely cause a shift in society due to a shortage of scientific workforce to defend the next pandemic incursion as published in the article (*RSC advances, (2024), 14, 37, 26897–26910*). A deeper understanding of quantum-coherent nanoscience and strategies for tackling technological obstacles helps to clarify the lab-on-a-chip path forward for the next wave of innovations and breakthroughs for creation of new technologies to delay or mitigate multimorbidity in aging world.

Bangkok, Thailand   Prof. Surachate Kalasin

Preface

This textbook started out as research publications and lecture notes that were given over several years to introduce developments in nanoscience and nanotechnology that drastically changed industries all over the world. Since the beginning of this decade, the industrialized world has witnessed an expansion of the nanoscale markets, indicating a revival of technological advancement. This textbook was written with the intention of introducing advanced graduate engineers, applied physicists, applied chemists, and other aspiring graduate students to the principles of nanomaterials, the function of sensor materials, and cutting-edge lab-on-a-chip technologies for decentralized healthcare and precision medicine. Nanotechnology-based products are steadily becoming more and more commonplace in daily life. Examples include smartphones, makeup, COVID-19 antigen testing kits, portable blood glucose monitors, building paints, and film coatings. Beneficially, a growing number of interdisciplinary fields are focusing on nanotechnology, including nanoelectronics, nanobiotechnology, nanomaterials, nanomedicine, and nanorobots, to advance lab-on-a-chip technologies.

Purposely, eight chapters comprise the book's content, which covers everything from the basics to modern developments and technological applications. The chapter contents include, but are not limited to, key concepts in the nanoscopic world, the existence of nanomaterials, solid-state physics, modern electrochemistry, nanomaterial synthesis, nanomaterial properties, material science for crystal structure polymer structures, tools in nanotechnology for sensors, DNA nanotechnology, and nanomaterial applications for precision medicine, the advancement of telemedicine, wearable sensors, and micro and nanorobots. Chapter 1 explores the evolution of nanoscience over time and discusses the nature of carbon, a fundamental material on Earth. Chapter 2 provides an understanding of the nanoscopic properties of solids at the fundamental level. It also focuses on the overall fundamentals of nanomaterials, quantum dots, and energy band theories. Chapter 3 introduces modern electrochemistry as a driving force for advancing lab-on-a-chip sensors. Chapter 4 also introduces tools in nanotechnology for developing sensors at nanoscopic and atomic levels, such as transmission electron microscopes (TEMs) and atomic force microscopes (AFMs).

A thorough understanding of how these tools operate is important to initiate any analysis of the spectra, morphology, and topology of nanostructures exploited in sensors. Chapter 5 introduces novel syntheses that are used to develop sensing materials for modern sensors. Multidisciplinary nanotechnology often involves DNA fabrication or motifs, which can detect potential biomarkers for disease identification, thereby creating more advanced sensing platforms. Today's healthcare systems are primarily overactive and overcrowded with patients. Upon the onset of symptoms, patients seek medical attention from doctors, who then treat and observe them passively, without actively seeking medical consultation. The DNA nanotechnology described in Chap. 6 takes molecular information out of its biological surroundings and uses it to piece together structural patterns. Chapter 7 discusses the advancement of healthcare using wearable sensors and lab-on-a-chip devices. Chapter 8 delves into the use of miniature machines, such as micro and nanorobots, for sensing applications. Imagine a future in which robots assist our elderly grandparents, operate cars, and teach children. These futuristic possibilities of miniaturized robots can be achieved. By going over the working principles of the nanotechnological approaches, the textbook investigates the present and potential uses of these materials in sensing and medicine.

In the end, it is a perfect manual for beginners, graduate students at any level, and advanced learners to help them quickly become acquainted with the advancements in nanoscience and to provide them a thorough understanding of the latest advancements and breakthroughs in lab-on-a-chip sensors.

Bangkok, Thailand   Prof. Surachate Kalasin

Acknowledgements Firstly, I would like to acknowledge the extraordinary debt I owe to the professors at the University of Massachusetts at Amherst for teaching me technological skills. I am grateful to my family and colleagues for their support and encouragement. I also thank King Mongkut's University of Technology Thonburi and the Thai government for financing my research on sensors and lab-on-a-chip technologies. Most importantly, I am filled with gratitude to the Springer Nature team for their gracious assistance throughout the textbook's preparation and production process.

Competing Interests The author has no competing interests to declare that are relevant to the content of this manuscript.

Contents

1 Imprints of Nanoscopic World in Times 1
   1.1 Nanoscience in Times 1
   1.2 Nature of Carbon 5
   1.3 Existence of Nanomaterials 7
   1.4 Renaissance of Nanocomposites 9
   1.5 The Importance of Nanoscience for Lab-on-a-Chip Sensor
        Development 10
   References 10

**2 Fundamental of Nanomaterials to Quantum Dots and Energy
Band Theories** 13
   2.1 Crystal Structure 13
   2.2 Theory of Diffraction 15
   2.3 Energy Bands of Solid Crystals 18
   2.4 Molecular Orbital Theory 20
   2.5 Imperfections in Solids or Defects 21
   2.6 Quantum Confinement and Deep Traps 22
   References 24

3 Electrochemical Theory for Micro and Nanoscale Essentials 27
   3.1 Faraday's Law 27
   3.2 Process of Electron Transfer 28
   3.3 Electrode and Surface Potentials 32
   3.4 Foundation of Nernst Response 33
   3.5 Electrode Classification 35
   3.6 Electrochemical Cell 40
   3.7 Electrode Reaction 41
   3.8 Electrocatalysis 43
   References 44

4	**Tools in Nanotechnology for Sensors**	47
	4.1 Scanning Tunneling Microscope (STM)	47
	4.2 Atomic Force Microscopy (AFM)	49
	4.3 Dip-Pen Nanolithography (DPN)	52
	4.4 Transmission Electron Microscopy (TEM)	55
	4.5 X-Ray Photoelectron Spectroscopy (XPS)	56
	References	58
5	**Nanomaterial Synthesis and Characterization**	61
	5.1 Bottom-Up and Top-Down Synthesis	61
	5.2 Metal Fluoride Synthesis	64
	5.3 Ionic Liquid	65
	5.4 Transition Metal Dichalcogenides (TMDCs)	67
	5.5 Green Synthesis of Nanoparticles	68
	References	69
6	**DNA Nanotechnology**	71
	6.1 Role of DNA	71
	6.2 DNA Motifs	72
	6.3 Wireframe DNA Origami	73
	6.4 Engineering Functional DNA–Protein Conjugates	75
	6.5 Bioassays	76
	References	78
7	**Development of Wearable Sensors for Sensing Applications**	81
	7.1 Components of Biosensors	81
	7.2 Detection of Small Molecules and Biochemical Conditions	85
	7.3 Sensing Materials for Advanced Biosensors	86
	7.4 Kidney-Disease and Global Public Health Issues	90
	7.5 Wearable Sensors for Kidney Monitoring	90
	References	95
8	**Self-sensing Intelligent Micro and Nanorobots for Monitoring Systems**	97
	8.1 Origin of Driven Micro and Nanorobots	97
	8.2 Micro and Nanorobots Powered by Chemicals	99
	8.3 Medical Micro and Nanorobots in Precision Medicine	100
	8.4 Targeted Drug Delivery	101
	8.5 Inorganic Material Agents	103
	8.6 Image Guidance for Monitoring Systems	106
	8.7 Untethered Micro and Nanomachines for Remote Intelligent Sensing	108
	References	110

About the Author

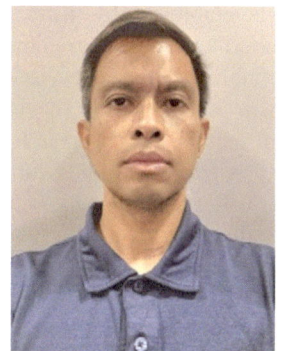

Surachate Kalasin received a Ph.D. degree in polymer physics from the University of Massachusetts at Amherst in the United States under the guidance of Professor Maria Santore. During the doctoral work, he completed a thesis, "Understanding and Exploiting Nanoscale Surface Heterogeneity for Particle and Cell Manipulation." The thesis included the design and development of selected nanopatterned biomimetic surfaces, regulated densities of functionalized chemical groups clustered together, and biosensor surfaces for medical applications. Moreover, he developed heterogeneous, patchy polymeric surfaces to separate and control the dynamic motions of cancer cells and organic particles by applying colloidal forces. Later, he completed his postdoctoral work at the department of polymer science and engineering at the same university. He dedicated most of his time to developing adhesives capable of sorting, collecting, or rejecting various types of biological cells, thereby strengthening his multidisciplinary skills. He also successfully completed the creation of biocompatible polymer surfaces for antimicrobial devices. Additionally, he was successful in creating artificial particles that specifically interacted with bacteria and phagocytes. With these achievements, he obtained several United States patents related to cancer cell sorting and separation.

Before attending the doctoral program, he completed a bachelor's degree in physics from Worcester Polytechnic Institute under the guidance of Prof. Stephen Pierson with the project "Study of fractional magnetic numbers in Quantum Cone Dots." Moreover, he finished another bachelor's degree in mathematics at the same university with the project "Mathematical Boundary Conditions of Electrochemical-Potential Models for Fuel-Cell Cathodes" under the supervision of Prof. Joseph Feribach and Prof. Bogdan Vernescu. Despite his interest in interdisciplinary subjects, he obtained master's degrees in electrical and computer engineering, along with physics, from the University of Massachusetts Amherst. His master's studies primarily concentrated on the fundamentals of heat dissipation in nanoelectronics computing, electronic transport in semiconductor devices, and band-structure calculation. He received the Peebles finalist award from the Adhesion Society for his graduate work.

Currently, Prof. Kalasin is employed by King's Mongkut University of Thonburi's graduate program in nanoscience and nanotechnology and supervises several graduate students. With a faculty role, he is also involved in all aspects of teaching and research, including curriculum design, program quality assurance for industrial standards, and acquiring research funds. His primary responsibility is to guide graduate students toward the completion of their Ph.D. degrees. As a result, he devotes a significant amount of time to teaching courses that require a deeper understanding of the subject matter while actively engaging with various technologies for research purposes. He also regularly serves as a professional reviewer in the scientific community. His primary areas of interest are wearable sensors for telemedicine, lab-on-a-chip technologies powered by artificial intelligence, biosensors, and computer simulation-based drug discovery. Finally, he believes that this textbook could motivate the younger generation to advance lab-on-a-chip technologies forward in an aging world and to mitigate or lessen multimorbidity in modern societies. Moreover, the technology also facilitates people living healthier with longer lives.

Chapter 1
Imprints of Nanoscopic World in Times

Abstract In a remarkable 1959 speech, physicist Richard Feynman predicted the rise of nanoscience in the modern world. The transcript talks about "plenty of room at the bottom," which came true in the 1980s as researchers created methods and instruments to examine and work with matter at the atomic nanoscale, less than 100 nm. Advances in nanoscience and nanotechnology have simplified modern life in almost every scientific and technical arena. According to estimates, this technology will play an important part in addressing the world's energy demands with clean fuels that emit the minimal pollutants. Nanoscience has the potential to create miniaturized electronics for non-invasive health monitoring, which can improve human health and lifespan. This chapter begins with the significant phenomenon of ribosome assemblers, a ubiquitous presence in living things.

1.1 Nanoscience in Times

The ribosome is a spectacular nanoscale assembler and may be the most well-known biological example of such molecular machinery. The ribosome, as displayed in Fig. 1.1, performs as a natural assembler for synthesizing proteins by attaching amino acids in a precise order through messenger RNA codons [1]. It is a challenging task. The ribosome must precisely determine a particular transfer RNA from at least 60 potential candidates—a feat that requires nanoscale precision [2, 3]. A messenger RNA chain translates the DNA sequence that determines the amino acid order in a protein. Ribosomes bind messenger RNAs and use their sequences to determine the proper amino acid order to produce a certain protein. Once the ribosome has chosen which amino acids to transport, the Transfer RNA (tRNA) functions in tandem with the messenger RNA to create amino acid chains via a codon stem loop. The three nucleotides that form the tRNA anticodon are paired with the three nucleotides present in the mRNA codon. Despite their vast disparities in size, the core structures of the different ribosomes are rather similar. Different tertiary structural motifs, such as coaxially stacking pseudoknots, arrange a significant portion of RNA. The RNA of the ribosome does all of its catalytic work, while the proteins that live on its surface

© The Author(s), under exclusive license to Springer Nature Singapore Pte Ltd. 2025
S. Kalasin, *Nanoscale Lab-on-a-Chip Sensors*, SpringerBriefs in Applied Sciences and Technology, https://doi.org/10.1007/978-981-96-5981-4_1

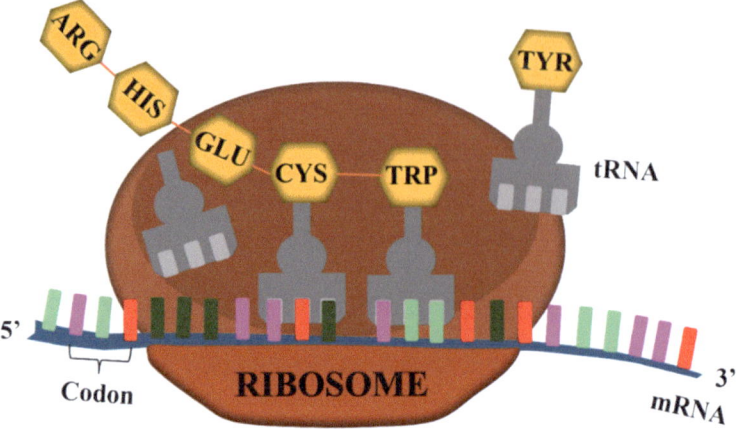

Fig. 1.1 Peptide synthesis by ribosome

appear to stabilize the structure [4, 5]. In this fashion, the ribosome machine can assemble proteins at a frequency of 20 Hz [6].

One more exquisite illustration of the function of nanostructures in everyday life is the process of photosynthesis, as depicted in Fig. 1.2a. During photosynthesis, green plants use light energy to convert water, carbon dioxide, and minerals into highly energetic and oxygen-rich organic molecules, such as sugars, cellulose, glycogen, and starches [7, 8]. As an important contributor to the creation and preservation of oxygen in the Earth's atmosphere, photosynthesis provides the majority of the biological energy required for complex life to exist on the planet. Everything that occurs in green plants takes place inside structures known as chloroplasts. Millions of these units reside in the leaves of plants, giving them their characteristic green hue. Thylakoids, several disks piled on top of one another, are found inside each chloroplast [9]. About 10,000 atoms and 200 colors make up each reaction center at the chloroplast. In addition, each reaction center has two extremely light-sensitive pigments that carry out the actual photon absorption [10, 11]. The fundamental units of the nervous system and brain comprise of neurons, or nerve cells, as they are depicted in Fig. 1.2b. The cells called neurons are responsible for taking in sensory data from the environment, controlling our muscles with motor commands, and converting and relaying electrical impulses at every step of the process [12, 13]. The human brain is without a doubt the most amazing piece of natural machinery we have yet to come across. It enables us to dream as well as think. With the advancement of computers, we have attempted to imitate or even surpass the human brain. But as we now know, computers are very different from human brains. Computer transistors are mostly fabricated with silicon, but brains are essentially with genome-wide patterns of carbon. The computer performs calculations and logical operations more quickly. On the other hand, the brain is more adept at creating new concepts and interpreting the outside world [14].

1.1 Nanoscience in Times

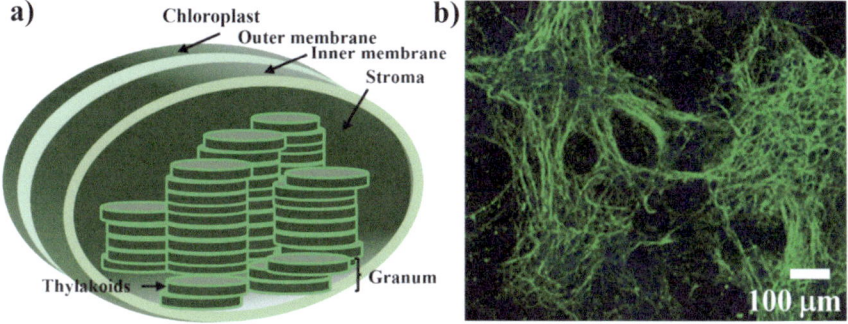

Fig. 1.2 **a** Photosynthesis and **b** network of neurons. Reproduced with permission from Ref. [13]. Copyright 2018 American Chemical Society

Many of the world's ancient empires, including the Parthenon in Greece, the pyramids in Egypt, and Machu Picchu in Peru, are renowned for their astounding large-scale engineering achievements. The artisans of those times were adept at engineering at the nanoscale, on the other extreme of the spectrum. However, even at the smallest level, ancient craftspeople also exerted influence over matters. Their work in the field of nanocomposites, by today's standards, was part of nanotechnology [15, 16]. It was realized that embedded nanoscale metallic particles within the glass produced the azure-sapphire color seen in Fig. 1.3a of numerous stained-glass windows from the Basilica's history. Richly colored stained glass, like to the metallic tones found in naturally painted nanoparticles in many glass windows, was greatly prized and highly valued by artisans, craftsmen, patrons, and laymen [17]. Natural nanomaterials are a vital component of life on Earth. They are found all over the environment and in natural species as well as being created as a result of geochemical processes [18]. In this regard, this essential geological factor is the natural process of nanoparticle synthesis. It has an evidence that oxidative DNA damage in human peripheral blood mononuclear cells has been shown to be induced by volcanic ash containing a high concentration of iron-bearing nanoparticles [19]. As seen in Fig. 1.3b, naturally generated nanoparticles originate from a variety of processes and are found in volcano, land, sea, air, and natural species [20]. Above the oceans, a large amount of atmospheric nanoparticles are produced. These are created by several sources and are referred to as marine aerosol. These fluctuate widely in size, but all are small enough to create an aerosol; like other aerosols, these are dominated by nanoparticles, with the majority of the nanofibers and nanosphere being about 15–55 nm in size, as measured as the number of particles per unit volume [21]. Many marine creatures, including fish, invertebrates, and marine mammals, have been shown to contain bioactive substances. Particles can bind together because glycoprotein macromolecules have a range of charges, including positive N-bonds, negative O, P, S, and C–O bonds [22].

It is important to note that, as shown in Fig. 1.4, human activity has produced certain nanoparticles that humans have been exposed to and breathed in since the

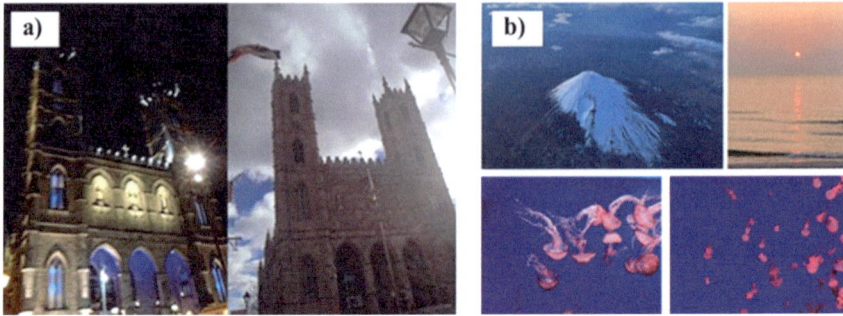

Fig. 1.3 a Window from Notre-Dame Basilica Church. The vaults are colored with deep blue of nanosized metal oxide added to the glass during the fusion process. **b** Nanoparticles abound everywhere

advent of fire. Life on our planet originated and has evolved in a dense cloud of naturally occurring nanoparticles. Furthermore, industrial processes utilizing specific nanoparticles as well as carbon black have existed since the dawn of human civilization. As an outcome, the skin, lungs, and intestines are the three main body surfaces that come into a contact with the nanomaterial waste product from the environment [23]. Many cultures have been using therapeutic aerosols and vapors for thousands of years; in fact, the use of therapeutic aerosols predates the use of vapors by at least 4000 years. In the late nineteenth and early twentieth centuries, numerous different nebulizers, including asthma cigarettes with stramonium and powders, were created, and attempts were made to provide several drugs by aerosol [24]. Worldwide, one of the main environmental risk factors for cardiovascular mortality and morbidity is exposure to particle air pollution. In particular, the highly bio-reactive proportion of nanosized particles found in urban air can enter almost every organ and is highly permeable [25].

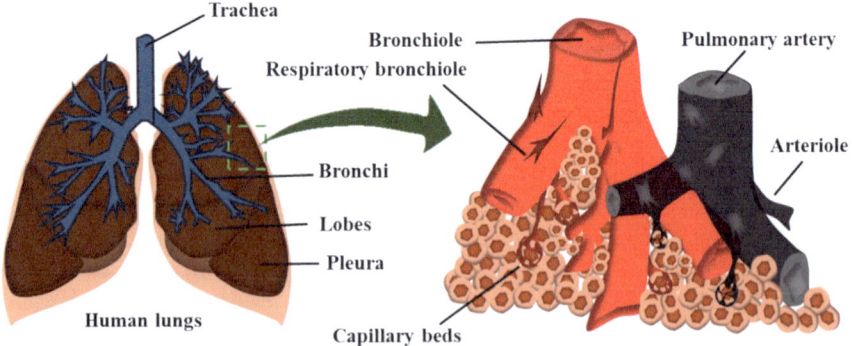

Fig. 1.4 Nano-aerosols in cardiovascular system

1.2 Nature of Carbon

All living creatures and their products include a significant amount of carbon. Carbon is also vital for nanotechnology research and applications. Carbon is a nonmetallic element found in living creatures' cells [26]. It is noteworthy to acknowledge that carbon constitutes approximately 19% of the mass of the human body and plays a crucial role in the synthesis of lipids, proteins, nucleic acids, and carbohydrates [27]. Figure 1.5a depicts a variety of low-dimensional carbon allotropes, such as graphene, fullerene, carbon nanotubes (CNTs), carbon fiber, and carbon black (CB). Both graphene and carbon nanotubes (CNTs) possess unique properties. Based on their atomic structure, CNTs are classified as metallic or semiconducting [28]. Carbon nanotubes (CNTs) are one of the most durable materials in one dimension and have excellent mechanic properties. Furthermore, fullerenes are allotrope carbon atoms that are made up of a closed or partially closed mesh with fused rings and are formed by alkane and alkene bonds. When n is the number of carbon atoms, the empirical formula C_n is usually used to define these fullerenes having a closed mesh structure. The values of n can sometimes be more than one isomer. Eiji Osawa historically predicted that C_{60} will exist in 1970. He noticed that this chemical form could exist and that polycyclic aromatic corannulene was a subset of a football. Harold Kroto and a team of scientists created the fullerene later in 1985 by vaporizing carbon under helium conditions. Individual peaks obtained from the mass spectrum were more likely to represent C_{60} or C_{70}, products with precise masses of sixty or seventy [29]. As was previously noted, carbon-based nanomaterials have been used in a wide range of industries, including electronics manufacturing, adhesive, fuel cell components, sensors, textiles, drug delivery, energy storage, packaging, and etc. [30, 31].

Over the past 50 years, diamond has undergone extensive examination, yielding remarkably detailed information on the interior of the earth. Although certain crystals are found at short depths, diamond is recognized as the sole material sample found in the very deep mantle, with depths exceeding 800 km. Diamond possesses the noteworthy capacity to monitor the mobility of carbon in the deep mantle. Diamond

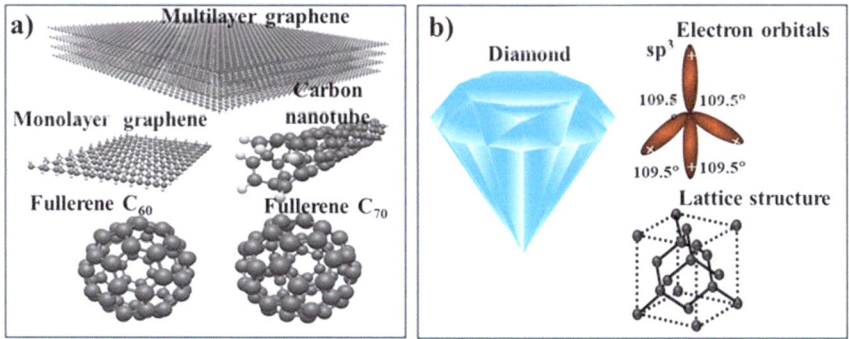

Fig. 1.5 **a** Different kinds of carbon nanostructures. **b** Electronic structure of diamond

thus has a remarkable place in every facet of the carbon cycle on Earth [26]. For instance, Macle diamond is twinned octahedra, and they are the most common type of octahedra found in diamonds [27]. Given in Fig. 1.5b, there are four electrons in each carbon atom that produce bonding with nearest neighbors and in diamond, the charge distribution associated with these electrons forms a tetrahedral structure (sp^3 orbitals) around each atom. The atoms thus come together in a tetrahedral arrangement. Furthermore, as seen by an SEM picture of unprocessed graphite scrap in Fig. 1.6a–c, graphite is formed by the multilayers of carbon atoms grouped into hexagonal rings with six carbon elements. The allotrope of carbon possesses sp^2 hybridization. Each carbon atom can form bonds with three other atoms in the sp^2 molecular orbital model. Neighboring carbon atoms in this bonding arrangement have 120° angles. The bonding carbons in graphite create three evenly spaced lobes with a plane charge distribution. Thus, the hexagonal arrangement of carbon atoms in graphite results in strong bonds within the sheets, but weak bonds between them. Conversely, when one rolls the sheet into a tube, preserving only the hexagonal carbon atom rings, we create a stable carbon structure known as a carbon nanotube, as illustrated in Fig. 1.7a, b. Iijima originally published images of carbon nanotubes captured with an electron microscope in 1991 [32]. These are carbon nanotubes with two to seven walls, or multiple-wall MWNTs. Starting with an infinite graphene sheet, families of tubes can be created by rolling up to link at atoms while maintaining the hexagonal lattice by cutting rectangles at various angles, such as (1)–(4).

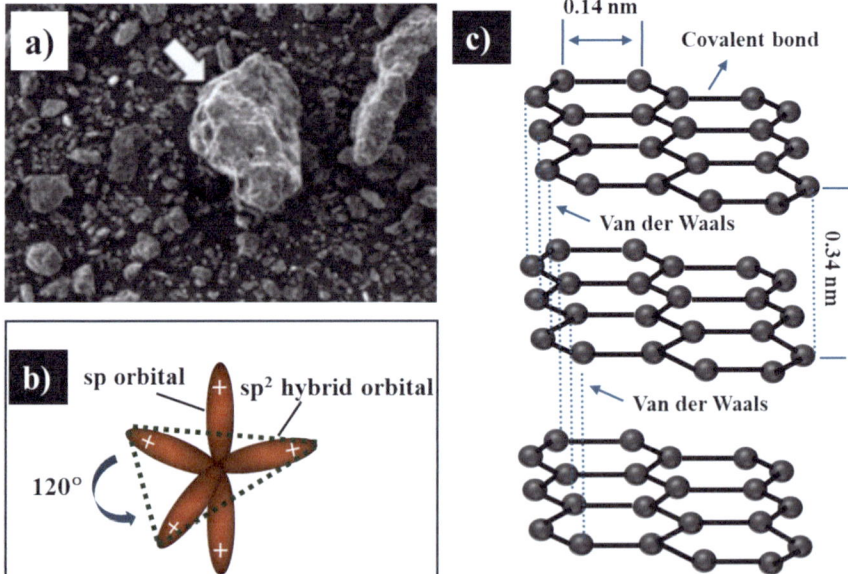

Fig. 1.6 **a** SEM image of raw graphite scrap at 100× magnification. Adopted with permission from Ref. [33]. Copyright 2023 CC-BY 4.0 American Chemical Society. **b** Orbital and **c** molecular structures of graphite

1.3 Existence of Nanomaterials 7

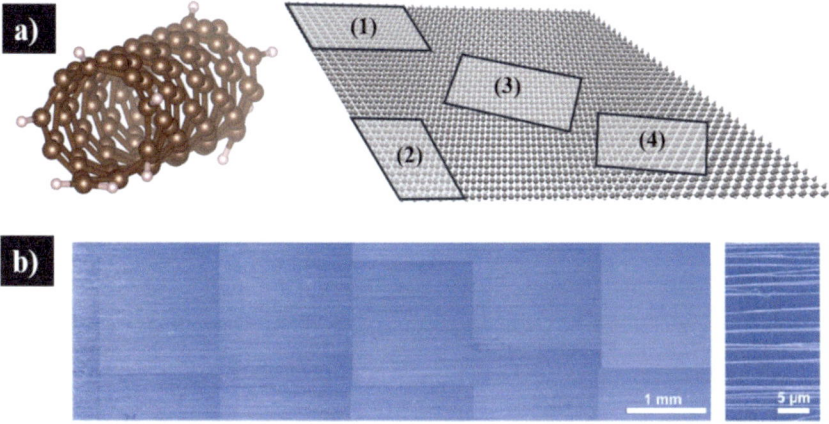

Fig. 1.7 Rolling of carbon nanotubes in different angles and SEM images of the high-density ultralong CNT array. Adopted with permission from Ref. [34]. Copyright 2023 American Chemical Society

1.3 Existence of Nanomaterials

Nanomaterials, sometimes known as nanoparticles or nanopowders, are finding their way into more and more aspects of daily life. Nanomaterials can be designed, engineered, constructed, or synthesized to have notable physicochemical characteristics at one or more dimensional scales, often between 1 and 100 nm [35]. Based on their dimensions, shape, composition, homogeneity, and aggregation, nanomaterials are typically categorized. As shown in Fig. 1.8a–d, they can be roughly divided into the following four categories based on dimensionality. Nanomaterials with controlled designs can have remarkably significant active surfaces. Nanomaterials that exhibit exclusive conductive, mechanical, optical, electrical, magnetic, reactive, and rapid catalytic properties that deviate markedly from bulk materials can be produced and synthesized [36]. For example, Au solution appears yellow in bulk but turns purple or red at the nanoscale. One can modify the properties of the nanomaterial by changing its size [37].

Due to their advantageous electronic structures for obtaining high catalytic activity and/or selectivity for a wide spectrum of processes, nanocrystals constructed of Pd, Pt, Rh, Ru, Ir, Ag, and Au, as well as mixtures of those given in Table 1.1, have attracted a lot of attention during the past several decades. Nanoscale colloidal metals are of interest in many disciplines. Methods for their preparation and chemical applications are of primary focus of chemists. The intense colors of colloidal gold metal, also known as gold sols, attracted attention since they may be employed as pigments in ceramics or glass. Because of the significant electrical structure of the nanosized metal particles and their extraordinarily huge surface areas, the number of possible uses for these colloidal particles is expanding quickly. Carbon quantum dots (QDs) as given in Fig. 1.9 are zero-dimensional discrete, quasi-spherical with a

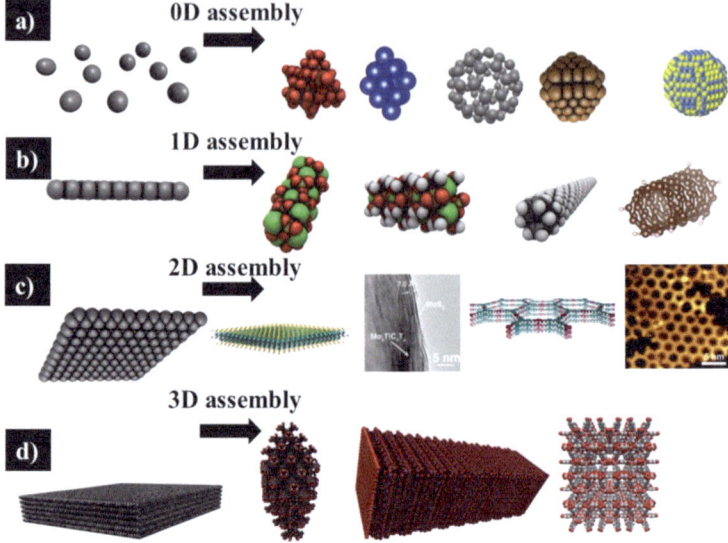

Fig. 1.8 Nanomaterial categories with **a** zero-dimensional (0D) fabrication. Reproduced with permission from Ref. [38]. Copyright 2021 American Chemical Society. **b** One-dimensional (1D) fabrication, **c** two-dimensional (2D) fabrication. Reproduced with permission from Refs. [39, 40]. Copyrights 2020 and 2021 American Chemical Society. **d** Three-dimensional (3D) fabrication

diameter of less than 10 nm. They have drawn a lot of attention to the development of nanotechnology. Two varieties of QD displays including color-converting and direct electrical-driving construct have been created in response to these advantages of this quasi-spherical material. Recently, to improve the contrast ratio and viewing angle of the displays even further, a patterned QD color-converting layer has also been developed to replace the traditional color filter [41].

Table 1.1 Examples of nanometals

Nanometals	Examples
Metals and alloys	Ti, Ti–C, Ti-transition metals (Fe, Cr, Ni, Cu) alloy; Fe-transition metal (Co, Ni, Mn, Cu, Nb) alloy; Fe–Cu–La–V–Mo alloy; Al, Al-transition metal (Zr, Zn, Cu, Ni) alloy; Mg, Mg–(Cr, Mn, Fe, Co, Rh, W, Ir) alloy
Nobel nanocrystal	Ru, Rh, Pd, Ag, Os, Ir, Pt, and Au

Fig. 1.9 Carbon dots (CDs) with tunable photoluminescence. Adopted with permission from Ref. [42]. Copyright 2016 American Chemical Society

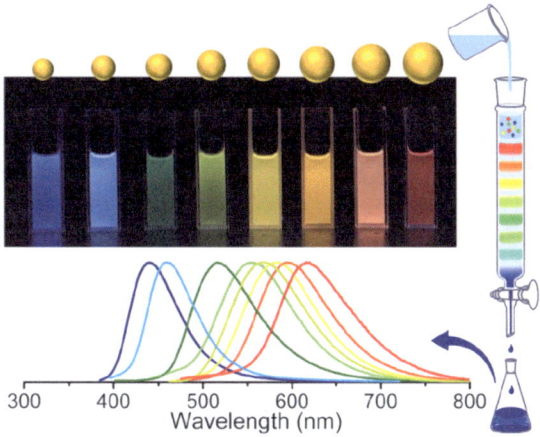

1.4 Renaissance of Nanocomposites

Composites are made with the intention of combining and utilizing the individual properties of their constituent parts [43]. As illustrated in Fig. 1.10, nature has perfected the process of creating strong composites by employing natural reagents and polymers such as carbohydrates, polysaccharide, lipids, and proteins. A nanocomposite material is composed of many phases, each of which has at least one, two, or three nanometer-sized dimensions. Nanocomposites present chances on entirely new dimensions for overcoming challenges in a variety of industries, including the pharmaceutical, food packaging, medical, electronics, and energy sectors. Around 3400 B.C., the ancient Mesopotamians made plywood by adhering wood strips at various angles, one of the oldest uses of composite material. The idea of "compositeness" has been around for a very long time [44]. Ancient plywood prepared by Mesopotamians gluing wood at different angles was found to give better properties than natural wood. Like carbon nanotubes in many aspects, cellulose nanofibrils are present in wood and other natural materials and have the potential to reinforce composites used in modern manufacturing.

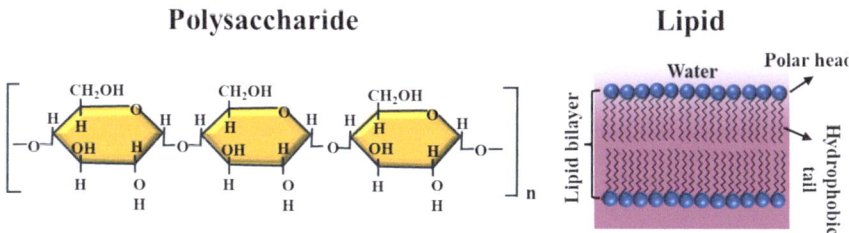

Fig. 1.10 Polysaccharide and lipids

1.5 The Importance of Nanoscience for Lab-on-a-Chip Sensor Development

Advances in nanotechnology simplify modern living. According to projections, this technology will play a significant role in helping to meet the world's energy needs with clean fuels that produce the least amount of air pollution. Miniaturized electronics for non-invasive health monitoring that can improve human health, and lifespan can also be advanced by nanoscience. Furthermore, by eliminating impurities through water filtration, nanotechnology can supply clean water to everyone on the planet. To boost crop productivity and reduce the need for pesticides, nanotechnology can provide innovative agrochemical agents and delivery methods. Significantly, essential characteristics such as size, shape, surface charge, surface area, surface porosity, composition, and structure are necessary for nanoscience to function effectively. Since many nanomaterials are heterogeneous and require in-depth information on crystal structure, energy band gaps, chemical functionalization, and grain boundaries, understanding the structure at the nanoscale is crucial. Modifications at the nanoscale can give rise to unique characteristics that differ from their macroscopic manifestations. The physical, chemical, optical, or mechanical constitution of a material might vary because of changes in surface area. We can make materials stronger, more resilient, or more conductive than their real-world equivalents. It appears that nanotechnology has multiple uses in several fields. As nanoscience progresses toward lab-on-a-chip sensors, wearable health monitoring sensors with point-of-care applications for today's aging society are emerging. Since nanotechnology is an essential part of modern life, nanoscience serves as a crucial multidisciplinary subject to develop futuristic materials for diverse technologies.

References

1. T.A. Steitz, A structural understanding of the dynamic ribosome machine. Nat. Rev. Mol. Cell Biol. **9**(3), 242–253 (2008)
2. S. Mann, Life as a nanoscale phenomenon. Angew. Chem. Int. Ed. **47**(29), 5306–5320 (2008)
3. S. Klinge, J.L. Woolford Jr., Ribosome assembly coming into focus. Nat. Rev. Mol. Cell Biol. **20**(2), 116–131 (2019)
4. J.A. Lake, The ribosome. Sci. Am. **245**(2), 84–97 (1981)
5. A.J. Samelson et al., Quantitative determination of ribosome nascent chain stability. Proc. Natl. Acad. Sci. **113**(47), 13402–13407 (2016)
6. S. Yan et al., Ribosome excursions during mRNA translocation mediate broad branching of frameshift pathways. Cell **160**(5), 870–881 (2015)
7. D.B. Knaff, D.I. Arnon, A concept of three light reactions in photosynthesis by green plants. Proc. Natl. Acad. Sci. **64**(2), 715–722 (1969)
8. T. Lawson, A.L. Milliken, Photosynthesis—beyond the leaf. New Phytol. **238**(1), 55–61 (2023)
9. J.F. Allen, J. Forsberg, Molecular recognition in thylakoid structure and function. Trends Plant Sci. **6**(7), 317–326 (2001)
10. U. Heber, H. Heldt, The chloroplast envelope: structure, function, and role in leaf metabolism. Annu. Rev. Plant Physiol. **32**(1), 139–168 (1981)

11. D.B. Stern, M. Goldschmidt-Clermont, M.R. Hanson, Chloroplast RNA metabolism. Annu. Rev. Plant Biol. **61**, 125–155 (2010)
12. H.R. Heekeren, S. Marrett, L.G. Ungerleider, The neural systems that mediate human perceptual decision making. Nat. Rev. Neurosci. **9**(6), 467–479 (2008)
13. W.L. Cantley et al., Functional and sustainable 3D human neural network models from pluripotent stem cells. ACS Biomater. Sci. Eng. **4**(12), 4278–4288 (2018)
14. M.M. Waldrop, Smart connections: computer chips inspired by human neurons can do more with less power. Nature **503**(7474), 22–25 (2013)
15. V.P. Sinoorkar, The Maya's knowledge of nanotechnology, in *History of Nanotechnology: From Pre-historic to Modern Times* (2019), pp. 91–111
16. N. Patkar, M. Sharan, Did nanotechnology flourish during the Roman Empire and Medieval periods?, in *History of Nanotechnology: From Pre-historic to Modern Times* (2019), pp. 113–140
17. M.G. Norton, Gold—the material of empire, in *Ten Materials That Shaped Our World* (Springer, 2021), pp. 65–85
18. M.F. Hochella Jr. et al., Natural, incidental, and engineered nanomaterials and their impacts on the earth system. Science **363**(6434), eaau8299 (2019)
19. C.J. Horwell, Grain-size analysis of volcanic ash for the rapid assessment of respiratory health hazard. J. Environ. Monit. **9**(10), 1107–1115 (2007)
20. I. Letchumanan, S.C. Gopinath, M.M. Arshad, Natural resources for nanoparticle synthesis, in *Nanoparticles in Analytical and Medical Devices* (Elsevier, 2021), pp. 45–57
21. B. Graca et al., Origin and fate of nanoparticles in marine water—preliminary results. Chemosphere **206**, 359–368 (2018)
22. A. Patwa et al., Accumulation of nanoparticles in "jellyfish" mucus: a bio-inspired route to decontamination of nano-waste. Sci. Rep. **5**(1), 11387 (2015)
23. A. Malakar et al., Nanomaterials in the environment, human exposure pathway, and health effects: a review. Sci. Total Environ. **759**, 143470 (2021)
24. F. Lavorini, F. Buttini, O.S. Usmani, 100 years of drug delivery to the lungs, in *Concepts and Principles of Pharmacology: 100 Years of the Handbook of Experimental Pharmacology* (2019), pp. 143–15
25. A. von Mikecz, T. Schikowski, Effects of airborne nanoparticles on the nervous system: amyloid protein aggregation, neurodegeneration and neurodegenerative diseases. Nanomaterials **10**(7), 1349 (2020)
26. Ö. Aktaş, F. Çeçen, Bioregeneration of activated carbon: a review. Int. Biodeterior. Biodegrad. **59**(4), 257–272 (2007)
27. R.A. Berner, The long-term carbon cycle, fossil fuels and atmospheric composition. Nature **426**(6964), 323–326 (2003)
28. M.-M. Titirici et al., Sustainable carbon materials. Chem. Soc. Rev. **44**(1), 250–290 (2015)
29. H. Kroto, The first predictions in the buckminsterfullerene crystal ball. Fullerenes Nanotubes Carbon Nanostruct. **2**(4), 333–342 (1994)
30. T. Belin, F. Epron, Characterization methods of carbon nanotubes: a review. Mater. Sci. Eng. B **119**(2), 105–118 (2005)
31. D. Tasis et al., Chemistry of carbon nanotubes. Chem. Rev. **106**(3), 1105–1136 (2006)
32. S. Iijima, P. Ajayan, T. Ichihashi, Growth model for carbon nanotubes. Phys. Rev. Lett. **69**(21), 3100 (1992)
33. C. Amnatsin et al., Ultrafine graphite scrap and carbon blocks prepared by high-solid-loading bead milling and conventional ball milling: a comparative assessment. ACS Omega **8**(50), 47919–47927 (2023)
34. Q. Jiang et al., Synthesis of ultralong carbon nanotubes with ultrahigh yields. Nano Lett. **23**(2), 523–532 (2023)
35. N. Baig, I. Kammakakam, W. Falath, Nanomaterials: a review of synthesis methods, properties, recent progress, and challenges. Mater. Adv. **2**(6), 1821–1871 (2021)
36. V.P. Sharma et al., Advance applications of nanomaterials: a review. Mater. Today Proc. **5**(2), 6376–6380 (2018)

37. L. Scarabelli et al., Plate-like colloidal metal nanoparticles. Chem. Rev. **123**(7), 3493–3542 (2023)
38. A.L. Efros, L.E. Brus, Nanocrystal quantum dots: from discovery to modern development. ACS Nano **15**(4), 6192–6210 (2021)
39. K.R.G. Lim et al., Rational design of two-dimensional transition metal carbide/nitride (MXene) hybrids and nanocomposites for catalytic energy storage and conversion. ACS Nano **14**(9), 10834–10864 (2020)
40. A.M. Evans et al., Two-dimensional polymers and polymerizations. Chem. Rev. **122**(1), 442–564 (2021)
41. E. Jang, H. Jang, Quantum dot light-emitting diodes. Chem. Rev. **123**(8), 4663–4692 (2023)
42. H. Ding et al., Full-color light-emitting carbon dots with a surface-state-controlled luminescence mechanism. ACS Nano **10**(1), 484–491 (2016)
43. R. Bogue, Nanocomposites: a review of technology and applications. Assem. Autom. **31**(2), 106–112 (2011)
44. P.A. Schroeder, G. Erickson, Kaolin: from ancient porcelains to nanocomposites. Elements **10**(3), 177–182 (2014)

Chapter 2
Fundamental of Nanomaterials to Quantum Dots and Energy Band Theories

Abstract Understanding the nanoscopic properties of solids at the fundamental nanoscale level is the main objective of this chapter. Considering the multiple particles present in solids, developing a nanoscopic explanation is a daunting endeavor. The classical or quantum–mechanical equations of motion of the particles are obviously unsolvable. Despite the enormous number of particles involved, it turns out that materials are frequently crystalline, with the atoms arranged on a regular lattice. This symmetry enables us to identify nanoscopic models. Despite the impression that these flawless crystals have no bearing on actual materials, this is untrue. Small crystalline grains make up the composition of many substances. Molecular organic crystals can be found in a wide range of material science fields. Most small-molecule medications are taken orally in the form of crystals. Molecular crystals are essential parts of pigments, insecticides, and fertilizers. They can be used as sensors, photovoltaic cells, field effect transistors, and light-emitting diodes.

2.1 Crystal Structure

An orderly arrangement of atoms, ions, molecules, or clusters in crystalline material is referred to as its crystal structure in crystallography. As depicted in Fig. 2.1a, the atoms in a crystal can be mathematically represented as points in a three-dimensional (3D) real space lattice. The real space unit vectors a, b, and c as well as the angles α, β, and γ can be denoted if these lattice points are organized in a periodic manner. The three linearly independent vectors a, b, and c that provide the basic primitive translation are generated by a particular vector as

$$T_n = n_1 a_1 + n_2 a_2 + n_3 a_3 \quad (2.1)$$

As seen in Fig. 2.1b, a crystal is a periodic pattern of atoms whose fundamental repeating unit is known as a unit cell. Primitive unit cell is a term used to describe the smallest unit cell. It serves to visually simplify the crystalline configurations that form solid in a periodic manner.

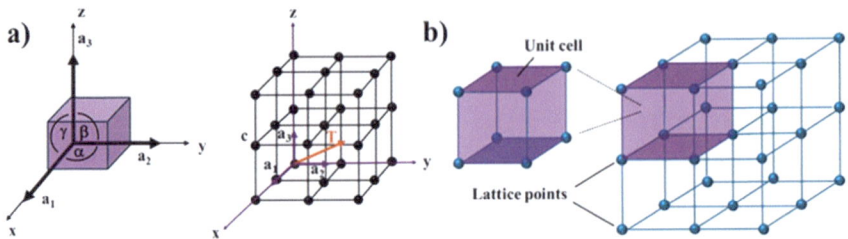

Fig. 2.1 a Crystal directions in real space. **b** Schematics of a unit cell with constructed lattice points

Every lattice site in a crystal lattice is at the border of a primitive unit cell and encloses the same surroundings [1]. Primitive unit cells lack uniqueness, just like primitive basis vectors do. The Wigner–Seitz cell, shown in Fig. 2.2, is by far the most often used primitive unit cell [2]. One can always locate a primitive cell with the full symmetry of the Bravais lattice, while the most common such choice is the Wigner–Seitz cell. This cell is placed in the first Brillouin zone in reciprocal space. It's also important to remember that lines linking each point in the lattice to every other point could be used to build the reciprocal cell. The seven crystal systems are the most evident way to categorize unit cells. As seen in Fig. 2.3, the Bravais lattice system additionally refers to extra configurational characteristics to split these seven systems into fourteen distinct Bravais lattices [3]. In addition, there are about two hundred and thirty kinds of space groups are available when accounting for the symmetry atom of the lattice structure [4]. The smallest primitive atoms that can repeat in an array to form the entire crystal, providing that they are symmetrically oriented. It should be noted that the letters a_1, a_2, and a_3 represent the unit cell dimensions, whereas the characters α_{12}, α_{23}, and α_{31} represent the crystalline angles within the unit cell axes. One of the most prevalent structures seen in binary ionic compounds is the Rocksalt or NaCl structure shown in Fig. 2.4. Larger chlorine atoms with an atomic radius of 181 pm occupy the face-centered cubic position in the structure of NaCl, while smaller sodium ions with an atomic radius of 102 pm are positioned in the octahedral space between them [5]. The two interpenetrating face-centered cubic (FCC) lattices make up the structure. Additionally, the sodium ions independently form an FCC lattice. Among the carbon allotropes, diamond is the hardest substance that has ever been discovered by humans [6]. It is a solid that is covalently bound and has two atoms at each lattice point in an FCC space lattice. The structure is sp^3-bonded and covalent. The packing density of this structure is quite low, at roughly 0.34 [7]. Two more crucial electrical elements that crystallize in the diamond cubic structure are silicon and germanium. A significant member of the II–VI group of semiconductor photocatalysts, zinc sulfide has exceptional physical and distinctive photocatalytic characteristics at the nanoscale. Its features include a significant exciton binding energy, a short Bohr radius, high activity under UV radiation, and a wide band gap energy (3.2–4.4 eV) [8]. ZnS can be found in two distinct crystalline forms, each with a distinct band gap energy: wurtzite and cubic sphalerite. Furthermore, the

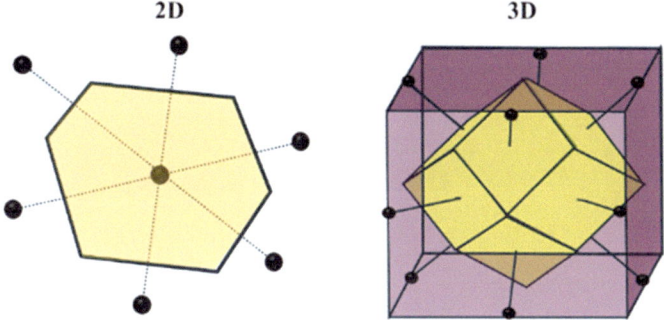

Fig. 2.2 Wigner–Seitz cells in two- and three-dimensional reciprocal lattices

application of ZnS quantum dots were employed as a fluorescence sensor to detect tetracycline quantitatively [9]. Dichalcogenides are chemical compounds made up of a chalcogen element such as sulfur and a transition metal such as molybdenum [10]. Silicon-based semiconductors devices were experiencing quantum and tunneling effects at the nanoscale level, MoS_2 shows positive and promising electrical and quantum characteristics while transitioning from bulk to two-dimensional structure [11].

2.2 Theory of Diffraction

Most analytical resolves of periodic structures, particularly in the theory of diffraction, rely heavily on the reciprocal lattice. Due to the Laue diffraction seen in Fig. 2.5a, the momentum difference between the incidence and diffracted X-rays of the crystal is a reciprocal vector in neutron, helium, and X-ray diffraction [12]. This procedure is used to assess the atomic orientation within the crystal [13]. When a plane wave scattering occurs by a crystal lattice, the Laue formulation can be defined as $\Delta k = k_r - k_i = G$, where $\Delta k, k_i$, and k_r stand for the scattering wave vector, incoming wave vector, and outgoing wave vector, respectively. The Laue postulation is based on the formulation that a lattice plane cannot be clearly identified by slicing crystals, and that the inclusion of ray reflection is not considered in the theory. Instead, the theory is predicated on the idea that each atom can radiate incident energy in all directions [14].

In the meantime, the Bragg assumption views the crystal as a collection of parallel ions that are spaced d apart from one another. Bragg diffraction, as shown by a crystal lattice in Fig. 2.5b, is a phenomenon that is comparable to thin film interference in that both have the same requirements for the refractive indices of the interfering and surrounding media. Furthermore, it is assumed that the target ion in any one plane will reflect the rays for there to be a strong peak. The parallel planes' reflected rays ought to cause constructive interference. The incident beam is a planar wave, as shown by

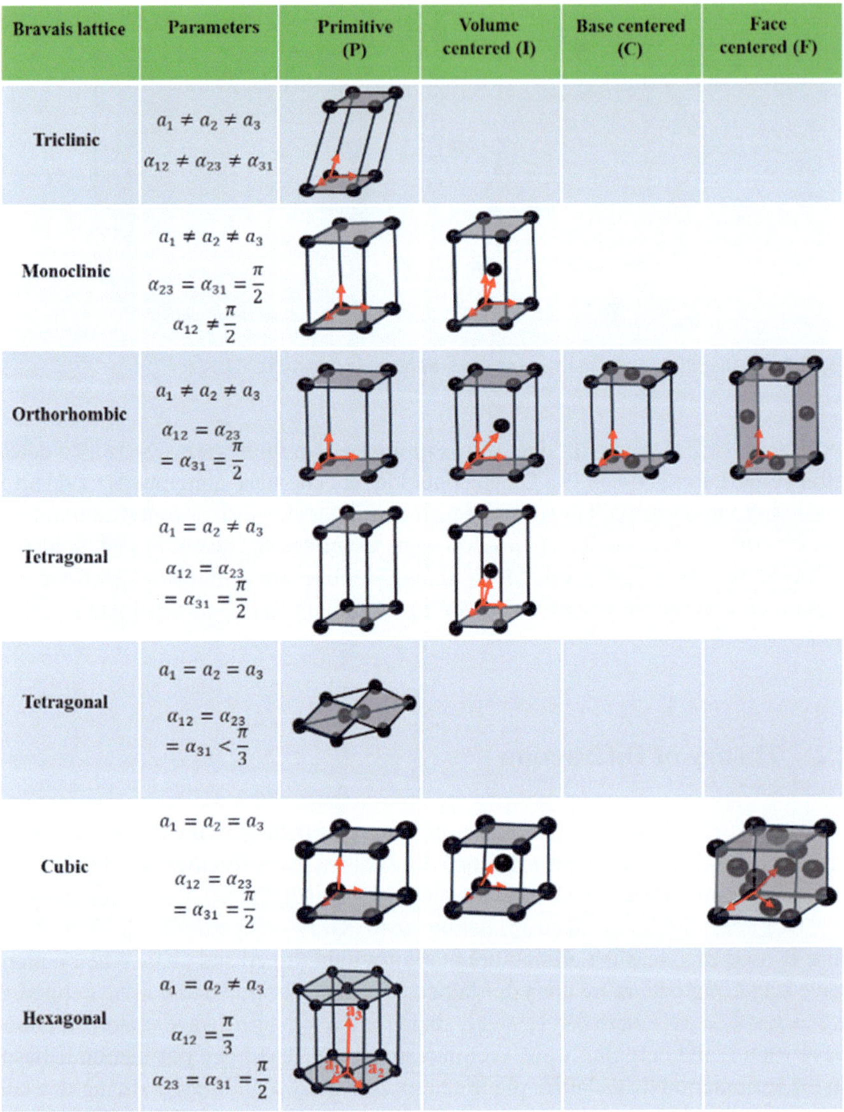

Bravais lattice	Parameters	Primitive (P)	Volume centered (I)	Base centered (C)	Face centered (F)
Triclinic	$a_1 \neq a_2 \neq a_3$ $\alpha_{12} \neq \alpha_{23} \neq \alpha_{31}$				
Monoclinic	$a_1 \neq a_2 \neq a_3$ $\alpha_{23} = \alpha_{31} = \frac{\pi}{2}$ $\alpha_{12} \neq \frac{\pi}{2}$				
Orthorhombic	$a_1 \neq a_2 \neq a_3$ $\alpha_{12} = \alpha_{23} = \alpha_{31} = \frac{\pi}{2}$				
Tetragonal	$a_1 = a_2 \neq a_3$ $\alpha_{12} = \alpha_{23} = \alpha_{31} = \frac{\pi}{2}$				
Tetragonal	$a_1 = a_2 = a_3$ $\alpha_{12} = \alpha_{23} = \alpha_{31} < \frac{\pi}{3}$				
Cubic	$a_1 = a_2 = a_3$ $\alpha_{12} = \alpha_{23} = \alpha_{31} = \frac{\pi}{2}$				
Hexagonal	$a_1 = a_2 \neq a_3$ $\alpha_{12} = \frac{\pi}{3}$ $\alpha_{23} = \alpha_{31} = \frac{\pi}{2}$				

Fig. 2.3 Configuration of different Bravais lattice

the graphical representation of Bragg diffraction, and the rays $\text{Ray}_{1,i}$ and $\text{Ray}_{2,i}$ are in phase before they reach the target ions. The additional distance that the ray $\text{Ray}_{2,f}$ must travel after reflection is given by the sum of $po' + p'o'$. It is evident from the illustration that $po' + p'o' = 2d \sin\theta$ for this distance. The additional distance of the second ray needs to be an integer number, $n\lambda$ to cause constructive interference. As a result, the relationship ends with $2d \sin\theta = n\lambda$ [15].

2.2 Theory of Diffraction

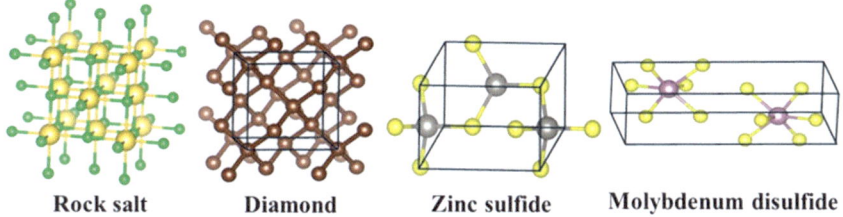

Fig. 2.4 Rock salt, diamond cubic and zinc sulfide, and MoS$_2$ structures

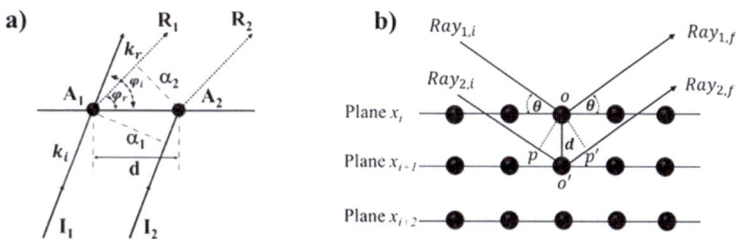

Fig. 2.5 **a** Graphical illustration of the Laue condition. **b** Bragg diffraction through a lattice of crystals

An Ewald sphere with radius k or the incident wave vector k are shown in a simple geometric construction of Ewald [16]. The reciprocal lattice vectors K, which are seen at the spherical borders, are responsible for the diffraction peaks. Theoretically, Fig. 2.6 shows the specific reciprocal lattice vector along with the Bragg reflected ray's wave vector, k'. As a result, the magnitude of the reciprocal lattice vector is expressed as follows,

$$|K| = 2\pi \frac{n}{d} \tag{2.2}$$

where n is the highest integral factors shared by the three Miller integers h, k, and l. The interplanar spacing distance for the set of Bravais lattice planes is, as usual, given by d. For instance, if the structure of a single nanocrystal sample is known, the Laue approach works best examining its orientation. Moreover, the spotting pattern created by the Bragg reflected rays will have the same symmetry if the incidence direction is along one of the crystal's symmetry axes [17].

Fig. 2.6 The crystal structure is determined by the geometric Ewald construction

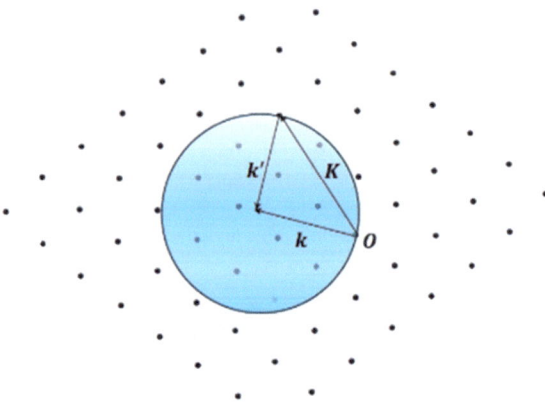

2.3 Energy Bands of Solid Crystals

We can gain useful insights into the thermal conductivity, electrical conductivity, magnetic susceptibility, impedance, and physicochemical dynamics of metallic crystals by using a free electron model. How the electrons react to an applied electric and magnetic field is a spectacular question about electrical resistance and conductivity. For example, each orbit of an isolated potassium atom has a specific amount of energy in it [18]. The energy levels of a pair of K atoms with molecular orbital combination, extracted in Fig. 2.7a. In this case, for the two potassium atoms, the surrounding atoms affect the energy state of the electrons in the outermost orbit. To create new molecular orbitals, bonding and antibonding energy levels require a combination of atomic orbitals. Sigma orbitals have bonding orbitals that are less energetic than antibonding orbitals. Accordingly, electrons closest to the nuclei of two atoms should be in the bonding orbitals, whereas electrons farthest from the nuclei would occupy in the antibonding orbitals. On the other hand, placing electrons within the antibonding orbitals will make the molecules less stable. Electrons will fill in accordance with the orbitals' energy levels. Higher energy orbitals will be occupied once the lower energy orbitals have been filled. The position of the energy levels as a function of interatomic spacing is visually shown in Fig. 2.7b. At very large atom separations, the energy scale is set to zero. The bonding level reaches its lowest energy for the separation a. The energy of the antibonding state is meaningless since only the bonding level is filled by two electrons; hence, the energy gain is greatest for the separation a. The atomic orbitals of the atoms overlap and change from the initial atomic orbitals when chemical bonds are established with them. The atoms' electrostatic interactions are what cause energy shifts. Explicitly, for the electron hopping between atoms to occur, the atoms need to be close enough. As shown in Fig. 2.8a, if N atoms with non-degenerated p energy levels are present in the crystal, then Np energy levels exist, each of which can hold two electrons and

2.3 Energy Bands of Solid Crystals

spin up or down. The energy level separations approach zero for very large numbers of atoms, $N \to \infty$. The energy differential that exists between the lowest and highest energy levels is a quasi-continuum that is referred to as a bandwidth or bandgap. For example, the figure of many several atoms illustrates, this energy band is half-filled with electrons (dark region) and half-empty (bright area) [19]. Each electron in the energy band diagrams of the various materials shown in Fig. 2.8b has a distinct energy level due to somewhat varied patterns of the surrounding charges. A continuous energy fluctuation is formed by these electron energy levels. The permitted energy zones are indicated by the vertical extent of the boxes. The regions filled with electrons, including valence and filled bands, are indicated by the shaded areas. The band with the lowest occupied energy levels with conduction electrons is known as the conduction band. All electrons are in the valence band when there is no external energy.

A material type classified as semimetal has very little overlap between the top of the valence band and the bottom of the conduction band [20]. Advanced electronics rely on inorganic semiconductors like germanium and silicon as its foundational materials for contemporary electronic gadgets. The semiconductors were found to have characteristics including light sensitivity, rectification, and negative temperature coefficient of resistance. The elemental semiconductors shown in Fig. 2.9a are

Two-atom system

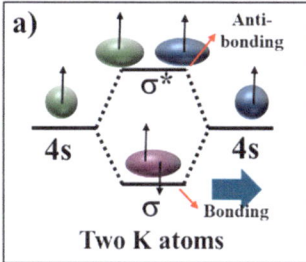
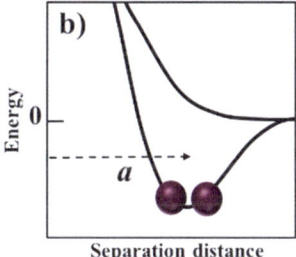

Fig. 2.7 **a** A system of two-potassium atoms with each electron configuration of $1s^2 2s^1 2p^6 3s^2 3p^6 4s^1$. **b** Energy level extracted with the separation distance of the two-atom system

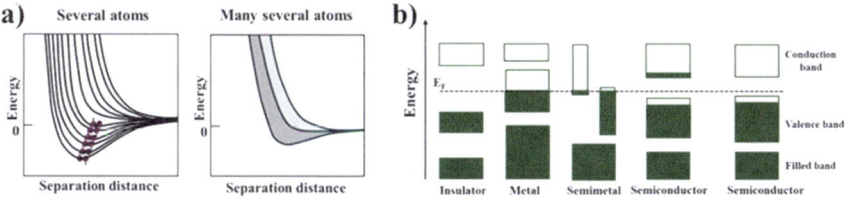

Fig. 2.8 **a** Many-atom energy levels and quasi-continuum atom energy band. **b** Energy band diagrams for semiconductors, semimetals, metals, and insulators are shown

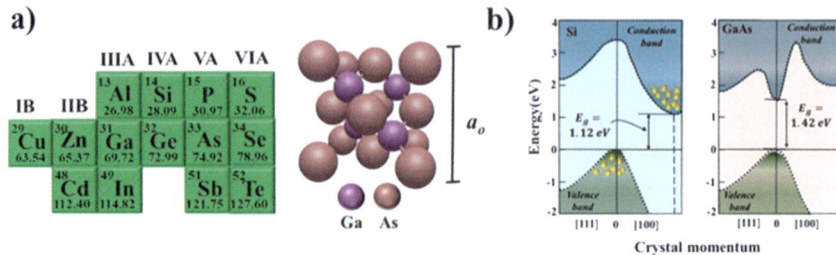

Fig. 2.9 a List of semiconductor elements. **b** Indirect and direct band gaps of silicon (Si) and gallium arsenide (GaAs), respectively

made up of individual atoms, like the ones found in columns IV, VI, and IVA of the periodic table, which are silicon (Si), Germanium (Ge), tin (Sn), and selenium (Se). By combining elements from groups IIIA and VA, IIB and VIA, compound semiconductors can be produced. Gallium arsenide (GaAs), gallium nitride (GaN), gallium phosphide (GaP), and indium antimonide (InSb) are a few examples of III–V compound semiconductors. Zinc oxide (ZnO), zinc sulfide (ZnS), cadmium sulfide (CdS), cadmium selenide (CdSe), and cadmium telluride (CdTe) are a few examples of II–VI compound semiconductors. For example, the band gap energies of Si, Ge, GaAs, and CdS are 1.11, 0.67, 1.43, and 2.42 eV, respectively [21, 22]. Another approach to classify semiconductors is as direct-band gap and indirect-band gap semiconductors, depending on the band structure. When the k-vectors or crystal momentum at the top of the valence band and the bottom of the conduction band occur at the same momentum value, this is known as a direct gap. If the crystal momentum is different, the material has an indirect gap, as seen in Fig. 2.9b. An electron can directly emit a photon if the crystal momentum of the electrons and holes is the same in the valence band and the conduction band. Since an electron needs to transfer momentum in crystal lattice through an intermediary state, a photon cannot be released in an indirect gap.

2.4 Molecular Orbital Theory

A three-dimensional area surrounding the nucleus where an electron is most likely to be discovered is called an atomic orbital. Electronic configuration, sometimes referred to as electronic structure or electron configuration, is the arrangement of electrons in orbitals surrounding the nucleus of an atom. In the quantum–mechanical model, the electronic configuration of an atom is given by the list of occupied orbitals, the number of electrons in each orbital, and the order in which they are filled [23]. Chemical properties are often different for atoms with different atomic numbers Z, and these properties also change periodically with Z. For example, alkaline metals with $Z = 3, 11, 19, 37, 55$, and 87 are lithium (Li), sodium (Na), potassium (K), francium (Fr), cesium (Cs), and rubidium (Ru). Their ionic characteristics are comparable [24]. For example, sodium can be melted into a crystal or ball that dissolves in water very quickly. Rubidium disappears almost rapidly with an explosion, melts very swiftly, and burns violently. The outer electrons of alkaline metals tend to be lost, forming positively charged ions. Consequently, there are more reactive metals the more readily these cations form [25]. The electron configurations of the first thirty many-electron elements each have a distinct set of occupied orbitals. The electron shells' spatial orbital extent is larger in higher-energy atoms. To be clear, Helium, neon, and argon atoms, which are closed-shell atoms at $Z = 2$, 10, and 18, have extremely closely bonded systems and need high ionization energies to ignite the outer-shell electrons [26]. With a structure consisting of a single loosely bound electron, the element lithium, sodium, and potassium at $Z = 3$, 11, and 19 primarily react with other electron-withdrawing atoms, especially the halogen elements by transferring their loosely outermost valence electron. $Z = 9$ and 17 atoms, fluorine, and chlorine, respectively, with a single electron vacancy at the outermost shell structure. Because they need an additional electron to favor a full electron shell, these atoms are reactive [27]. Therefore, one must understand the significance of quantum theory in relation to the Pauli exclusion principle. According to the rule, no two electrons can have the same electron spin or the same set number of quantum numbers.

2.5 Imperfections in Solids or Defects

Real crystals are imperfect as there are the presence of inherent defects. Imperfections have a significant impact on the characteristics of materials. Point defects are defined as flaws with a size restriction to about atomic dimensions, as illustrated in Fig. 2.10a. Point defects can be defined as impurities, like dopants. The point flaws that are present even in pristine material are referred to in this context as intrinsic point defects, interstitial, and vacancies. Moreover, pairings, also known as Frenkel pairs, can generate point defects in infinite crystals. The autonomous formation of interstitials and vacancies, as well as Schottky defects, is made possible by the existence of imperfect surfaces. A Schottky defect given in Fig. 2.10b is a particular form of point defect or imperfection in solids that develops when a vacant place in a crystal lattice is created as atoms or ions traveling from the crystal interior to its surface. Examples are KBr, $NaCl$, and KCl. The solid's density decreases as an equal number of cations and anions being absent from the lattice site. Furthermore, a point defect known as a Frenkel defect results when an atom or anion shifts from its initial location in the lattice structure to fill a void while occupying another interstitial position inside the solid crystal. Examples include ZnS and $AgBr$.

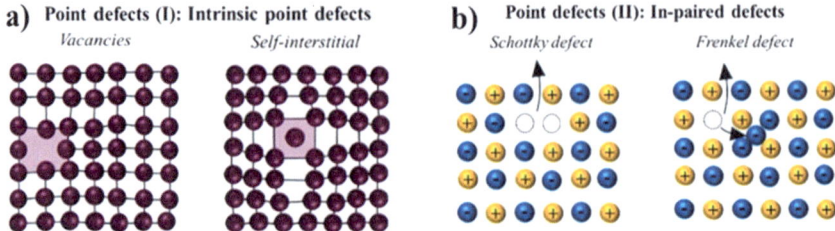

Fig. 2.10 **a** Intrinsic point defects: vacancies and self-interstitials. **b** Point defects: Schottky defect and Frenkel defect

2.6 Quantum Confinement and Deep Traps

Quantum dots are employed more often in a variety of applications in nanotechnology, from biosensors to the fundamental framework for quantum information processing and transmission. One of the most frequent issues in electrical devices is the requirement to address the positioning of a thin layer of nanostructure between two layers of a material with a higher bandgap. In the nanostructure, defects and traps are frequently quantum states. To explain this problem, Schrödinger's Eq. 2.3 describes the behavior of electron translocation in the quantum deep trap.

$$\hat{H}\varphi(x) = \left(\frac{\hat{p}^2}{2m} + V(x)\right)\varphi(x) = E\varphi(x) \tag{2.3}$$

Both defects and traps depend on quantum states, which are mathematically defined by the potential well shown in Fig. 2.11, and where the depth of the finite potential allows the trapped particle to escape. The potential space for this constant potential can be divided into three distinct zones, each having a constant potential. The localization of the trapped particle occurs in the limited well when its energy $E < V_0$. The Hamiltonian symmetry is even, this parity attains discrete quantum states, and the overall solution can be obtained with even or odd. In the region $-\frac{D}{2} \leq x \leq \frac{D}{2}$, where $V = 0$, the time independent Schrödinger equation reads:

$$\frac{d^2}{dx^2}\varphi(x) = -\frac{2mE}{\hbar^2}\varphi(x) = -k^2\varphi(x) \tag{2.4}$$

The obtained solutions take cos (kx) and sin (kx), where $k^2 = \frac{2mE}{\hbar^2}$. The region $|x| \geq \frac{D}{2}$ has a finite potential V_0. The time independent Schrödinger equation turns into

$$\frac{d^2}{dx^2}\varphi(x) = \frac{2m}{\hbar^2}(V_0 - E)\varphi(x) = \alpha^2\varphi(x) \tag{2.5}$$

and

2.6 Quantum Confinement and Deep Traps

Fig. 2.11 Diagram representing the finite potential well for a defect or quantum dot

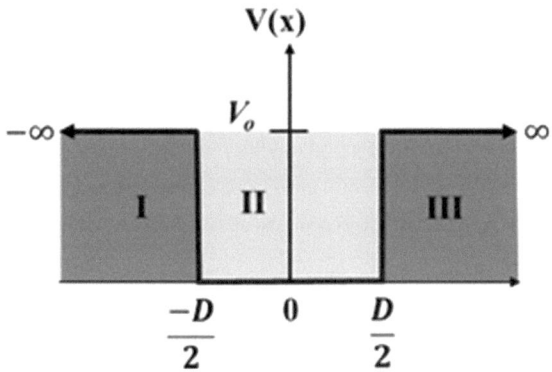

$$\alpha^2 = \frac{2m}{\hbar^2}(V_o - E) > 0$$

Specifically, the solution can be expressed in forms of exponential functions and is not zero as the outside potential is finite. One notices that $e^{-\alpha x} \to \infty$ as $x \to \infty$ and $e^{\alpha x} \to \infty$ as $x \to \infty$. These wave functions are divergent at the point $x = \pm \infty$. As the energies are quantized with quantum states, the solutions exhibit evident characteristics. In this instance, the particle can tunnel past the well barrier. There are several uses for this quantum tunneling in the advancement of semiconductor technology. For example, the chemical stability of the biosensors might be provided by metal oxide-based semiconductors, which would also improve the sensors' physicochemical interfacial qualities and their ability to form composite structures [28]. In addition, a potent method of measuring protein conformation in real time could be made possible by single-protein-coupled quantum mechanical tunneling (QMT) probes, which enable the capture and study of individual proteins in a defined tunneling gap. This would open the door to new kinds of single-molecule sensors and potentially sequencing platforms [29]. Under the Brus postulation with the theoretical expression given as,

$$E_{g(qd)} = E_{bulk} + \frac{h^2}{8R^2}\left(\frac{1}{m_e^*} + \frac{1}{m_h^*}\right) - \frac{1.786e^2}{4\pi \varepsilon_o \varepsilon_r R^2}, \tag{2.6}$$

where $E_{g(qd)}$ is a band gap energy of the quantum dot, R defines as a radius, m_e^*, and m_h^* are the effective mass of electron and hole, respectively. The quantum confinement phenomenon in silicon quantum dots can also be studied. It can be observed that the particle-in-a-well model predicts the size dependency in the simple models. Radius inversion determines the ground state energy in a proportionate manner. Consequently, the confinement energy falls but never reaches zero as one raises the size of the quantum dot [30].

References

1. H. Burzlaff et al., *Crystal Lattices* (2016)
2. P. Zou, R. Bader, A topological definition of a Wigner–Seitz cell and the atomic scattering factor. Acta Crystallogr. A **50**(6), 714–725 (1994)
3. L. Frevel, A systematic classification of crystal structures. Acta Crystallogr. B **41**(5), 304–310 (1985)
4. R.A. Evarestov, Space groups and crystalline structures, in *Quantum Chemistry of Solids: The LCAO First Principles Treatment of Crystals* (2007), pp. 7–46
5. R. Fuentes-Azcatl, M.C. Barbosa, Sodium chloride, NaCl/ϵ: new force field. J. Phys. Chem. B **120**(9), 2460–2470 (2016)
6. F. Herman, Electronic structure of the diamond crystal. Phys. Rev. **88**(5), 1210 (1952)
7. M. He et al., Colloidal diamond. Nature **585**(7826), 524–529 (2020)
8. R. Nomura et al., Useful and accessible organometallic precursors for preparation of zinc sulfide and indium sulfide thin films via solution pyrolysis (printing) method. Appl. Organomet. Chem. **4**(6), 607–610 (1990)
9. J. Zhang et al., Copper doped zinc sulfide quantum dots as ratiometric fluorescent probes for rapid and specific detection of tetracycline residues in milk. Anal. Chim. Acta **1216**, 339991 (2022)
10. Z.M. Wang, C. Chen, Z. Yu, MoS_2, in *PTCDA Hybrid Hetero* (2013)
11. V.P. Kumar, D.K. Panda, Next generation 2D material molybdenum disulfide (MoS_2): properties, applications and challenges. ECS J. Solid State Sci. Technol. **11**(3), 033012 (2022)
12. M. Eckert, Max von Laue and the discovery of X-ray diffraction in 1912. Ann. Phys. A83–A85 (2012)
13. D. Cruickshank, J. Helliwell, K. Moffat, Multiplicity distribution of reflections in Laue diffraction. Acta Crystallogr. A **43**(5), 656–674 (1987)
14. J. Zak, Lattice operators in crystals for Bravais and reciprocal vectors. Phys. Rev. B **12**(8), 3023 (1975)
15. C.G. Pope, X-ray diffraction and the Bragg equation. J. Chem. Educ. **74**(1), 129 (1997)
16. Y.L. Loh, The Ewald sphere construction for radiation, scattering, and diffraction. Am. J. Phys. **85**(4), 277–288 (2017)
17. P. Ewald, The principles of X-ray diffraction, in *Fifty Years of X-Ray Diffraction: Dedicated to the International Union of Crystallography on the Occasion of the Commemoration Meeting in Munich July 1962* (Springer, 1962)
18. J. Singleton, *Band Theory and Electronic Properties of Solids*, vol. 2 (OUP, Oxford, 2001)
19. G.W. Erickson, Energy levels of one-electron atoms. J. Phys. Chem. Ref. Data **6**(3), 831–870 (1977)
20. A. Kahn, Fermi level, work function and vacuum level. Mater. Horiz. **3**(1), 7–10 (2016)
21. M. Pohanka, J. Leuchter, Biosensors based on semiconductors, a review. Int. J. Electrochem. Sci. **12**(7), 6611–6621 (2017)
22. N. Chaniotakis, N. Sofikiti, Novel semiconductor materials for the development of chemical sensors and biosensors: a review. Anal. Chim. Acta **615**(1), 1–9 (2008)
23. A. Streitwieser, Molecular orbital theory for organic chemists, in *Pioneers of Quantum Chemistry* (ACS Publications, 2013), pp. 275–300
24. C.C. Addison, *The Chemistry of the Liquid Alkali Metals* (1984)
25. F.E. Beamish, *The Analytical Chemistry of the Noble Metals* (Elsevier, 2013)
26. G. Frenking et al., Light noble gas chemistry: structures, stabilities, and bonding of helium, neon, and argon compounds. J. Am. Chem. Soc. **112**(11), 4240–4256 (1990)
27. J.L. Magee, The mechanism of reactions involving excited electronic states the gaseous reactions of the alkali metals and halogens. J. Chem. Phys. **8**(9), 687–698 (1940)
28. I. Şerban, A. Enesca, Metal oxides-based semiconductors for biosensors applications. Front. Chem. **8**, 354 (2020)

29. L. Tang et al., Measuring conductance switching in single proteins using quantum tunneling. Sci. Adv. **8**(20), eabm8149 (2022)
30. S. Harry, M. Adekanmbi, Confinement energy of quantum dots and the Brus equation. Int. J. Res.-Granthaalayah **8**(11), 318–323 (2020)

Chapter 3
Electrochemical Theory for Micro and Nanoscale Essentials

Abstract The term "*modern electrochemistry*" has many different definitions after the invention of flexible electrodes. The broadest definition, which seeks to encompass all facets, can be described as science that strives to enhance human existence by utilizing both the bulk of electrolyte solutions and the phenomena that arise at the interfaces between metallic or semiconducting electrodes and electrolyte solutions. Understanding these phenomena can lead to the development and design of new systems or devices that can be used in both healthcare industry and more specialized applications. Examples of these include implanted electricity generators, pacemakers, self-powered electronics, smartphones, and disease detection devices. Faraday's law, the foundational principle of all electronic sensing devices, begins in this chapter.

3.1 Faraday's Law

The needed or released charge in a redox reaction is correlated with the number of species involved, according to Faraday's law. The law can be formed by considering the mass of the species, such as the mass of the deposited layer in the case of electrodeposition.

$$m = \frac{M}{zF}Q \tag{3.1}$$

According to this, m stands for mass, M for molar mass, z for valency, and Q for charge. Currently, the Faraday constant F is expressed in terms of the elementary charge q_e and the Avogadro constant N_A, whereas historically it was acquired empirically.

$$F = q_e N_A = 96,485 \text{ A s mol}^{-1} \tag{3.2}$$

Alternatively, Faraday's law can be expressed more succinctly by substituting the current density i for the charge and the species flux density j for the mass.

© The Author(s), under exclusive license to Springer Nature Singapore Pte Ltd. 2025
S. Kalasin, *Nanoscale Lab-on-a-Chip Sensors*, SpringerBriefs in Applied Sciences and Technology, https://doi.org/10.1007/978-981-96-5981-4_3

$$i = zFj \tag{3.3}$$

This version of Faraday's law enables the connection between a species concentration and a measured current in a diffusion-controlled situation, or it can be used to connect the current and the typical species flux into an electrode while simulating the electrode's microenvironment [1, 2].

3.2 Process of Electron Transfer

A redox process, or chemical reaction involving the transfer of electrons (oxidation or reduction), is known as a faradaic process. Consequently, a current resulting from a shift in redox state is known as a faradaic current. One typical misconception is to confuse capacitive currents with Faraday currents. A capacitive effect, on the other hand, is a non-faradaic process since it does not involve a change in redox state [3]. The prototypic n-electron processes are described using the fictive species O (oxidized form) and R (reduced form) in the explanation of electrochemical theory and methodology.

$$\text{Reduction: } O + ne^- \rightarrow R \tag{3.4}$$

$$\text{Oxidation: } R \rightarrow O + ne^- \tag{3.5}$$

Electron transport is a key concept to grasp while attempting to comprehend a Faradaic process. There are various facets as well as various abstraction levels. Each method ultimately aims to represent the complex issue using a straightforward number that corresponds to impacts that may be observed through experimentation. Knowing the assumptions made during the simplification is crucial from a practical standpoint. The energy diagram of an electrode for a metal reduces to the electron energy level. The molecular makeup of the ingredients must be considered in the solution. Given in Fig. 3.1, one must consider the lowest unoccupied molecule orbital (LUMO) and the highest occupied molecule orbital (HOMO) when examining the energy diagram. When an electron from the electrode fills the LUMO through a reduction process or departs the HOMO through an oxidation transition, this is known as electron transfer. When an electron is energetically favored while it is at a lower energy level, it can transit along the electrode and solution contact. Higher energy levels of the electrons in the electrode are necessary to drive the reduction reaction. The relationship between the energy ΔG and the electrical potential E is inverse.

$$\Delta G = -nFE \tag{3.6}$$

3.2 Process of Electron Transfer

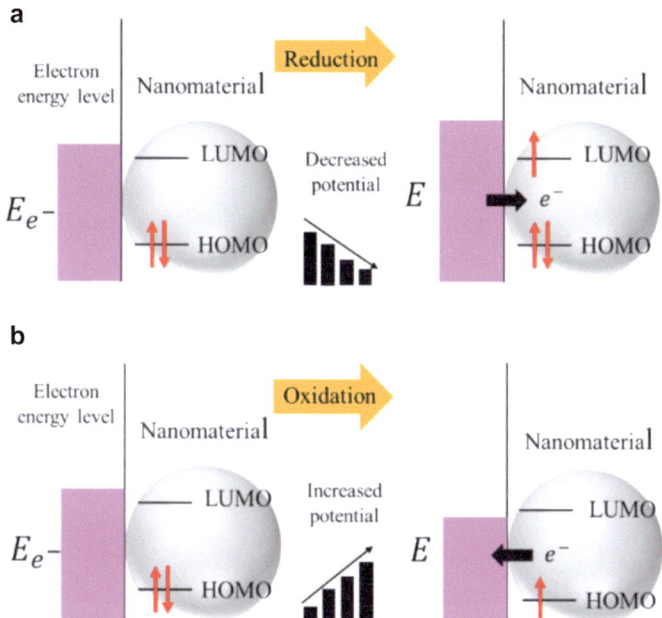

Fig. 3.1 Energy diagrams represent the situation at the electrode and solution interface: electron transport at a metal electrode. **a** Increasing the electrode potential (reducing the energy level of an electron) promotes oxidation, whereas **b** decreasing the electrode potential (increasing the energy level of an electron) promotes reduction

For instance, Gibbs free energy requires for the process of Au^{3+} reduction to metallic gold, one can write $-\Delta G° = 3FE°(Au^{3+}/Au)$. When the electrode's electrical potential is decreased and activating the reduction process, the electron energy level increases. This procedure, which entails the transfer of an electron from the electrode, needs to have enough energy and low enough potential to reduce any absorbed layers or nanomaterials. By increasing the electrode's electrical potential and lowering the electron's energy level, the oxidation process via electron transfer to the electrode is triggered [4]. Not only is there a relationship between energy and potential, but there are also various potential scales in physics and electrochemistry when it comes to a reference potential. In physics, the reference point is typically the vacuum level or the state of electron rest, whereas in electrochemistry, the standard metallic or gas potential is used. In Fig. 3.2, all potentials E denotes standard hydrogen energy (SHE). As a result, the H^+/H_2 reaction's standard potential is zero [5].

Nitrogen gas reacts with hydrogen gas to form ammonia through synthesis reaction. A further crucial characteristic is iron reduction (Fe^{3+}/Fe^{2+}) at 0.77 V [6]. Lithium reacts with hydrogen gas to produce lithium hydride, demonstrating both its great reactivity and vast potential for energy storage. The temperature-dependent

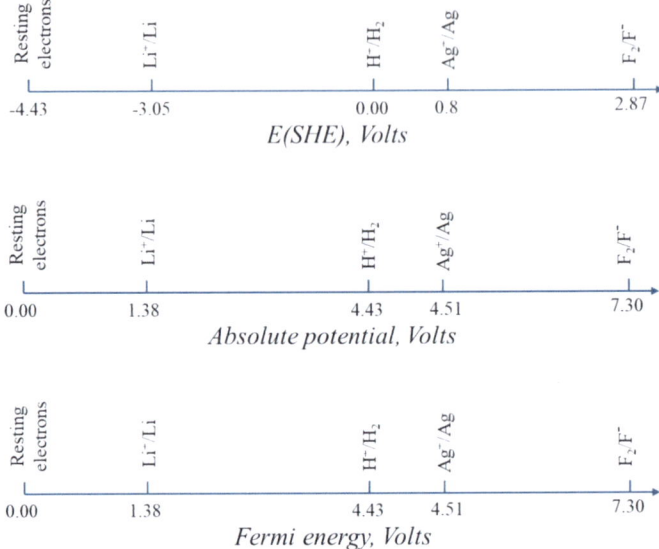

Fig. 3.2 Integrating physical and electrochemical scales. The energy scale and absolute potential are related to the resting energy of an electron. The values are provided at 25 °C

Fermi energy at the standard hydrogen electrode (SHE) determines the precise relationship between the scales. The value at 25 °C is − 4.44 V for resting energy of an electron [7]. When relating at various scales, additional effects might be involved with real electrodes. Even while electron transport is always necessary, an electrode reaction's rate-determining step could not always be it. Gazing beyond the electrode and solution contact, we must consider the approaches and departures of reduced and oxidized *species* from the interface. Figure 3.3 provides an overview of the several stages that comprise the oxidation and reduction processes. The reduced species (R) must get to the electrode's microenvironment from somewhere to the electrolyte interface. This mass transport is dominated by either migration, convection, or diffusion. Diffusion, in any event, controls the final distance in overcoming an interlayer to the electrode. The mass transfer from the bulk condition (species R) to the immediate vicinity of the electrode (species indicated as R^*) is represented by the mass transfer coefficient k_m. The R^* must find its way to the electrode surface where it absorbs as R^{**} before the electron transfer can occur. The k_{ads} is the corresponding adsorption rate constant. The oxidation reaction from R^{**} to O^{**} is quantified by the charge-transfer rate constant of oxidation, k_{ox}. Following this, O^{**} desorbs to O^* and exits the electrode's proximity to become finalized O species in the bulk. The solid oxidation process pathway is depicted in the figure. The reduction process, which follows the same steps but in the opposite order, is shown by the dashed lines. In this case, k_{re} would be the rate constant of reduction. In contrast to adsorption, the electron transfer is frequently the rate-determining step. As a result, the question of whether mass transfer or electron transfer limits the reaction is typically

3.2 Process of Electron Transfer

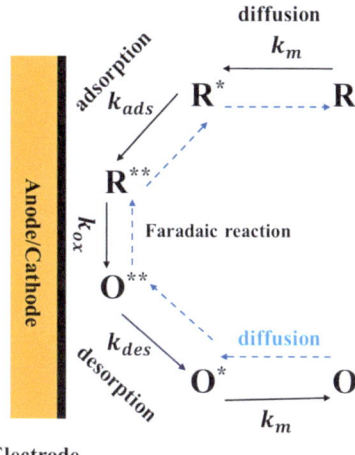

Fig. 3.3 Adsorption and desorption in a general electrode reaction. The stages of an oxidation reaction at the anode with the charge-transfer rate constant k_{ox} are shown by solid lines. The reduction reaction, for which the cathode would be the electrode, is depicted by the dashed lines. In this case, k_{re} would be a rate the charge-transfer rate constant

discussed. The charge transfer description is permitted in the latter scenario by the Butler–Volmer equation. The reaction rate v connects the species concentration c close to the electrode with the rate constant k for electrode reactions as defined by the Butler–Volmer equation.

$$v = kc \tag{3.7}$$

By using Faraday's law to relate the reaction rates to the anodic i_a and cathodic i_c current densities and an electron transfer factor α, the reaction rate may also be regarded as the species flux into or out of the electrode [8]. Thus,

$$i_a = nFk_0[R^*]\exp\frac{\alpha_a nF(E-E^0)}{RT} \tag{3.8}$$

$$i_c = -nFk_0[O^*]\exp\frac{\alpha_a nF(E-E^0)}{RT} \tag{3.9}$$

When the reduction and oxidation currents are combined, the overall current can be expressed as

$$i = i_a + i_c \tag{3.10}$$

The Butler–Volmer equation is now as follows

$$i = i_0 \cdot \left(\exp\frac{(1-\alpha)nF\eta}{RT} - \exp\frac{-\alpha nF\eta}{RT}\right) \tag{3.11}$$

In this case, the deviation of electrode potential η from equilibrium in this instance might be realized as the overpotential.

$$\eta = E - E^0 \tag{3.12}$$

Due to its resemblance in appearance, the Butler–Volmer equation is sometimes referred to as the diode equation of electrochemistry [9].

3.3 Electrode and Surface Potentials

A potential leap or drop may happen at the phase interface when two solid conductive materials with the same or opposing charges come into touch with one another or when the accumulation of charges on both sides forms a boundary between them. Most of the extremely thin layer of charge is centered on the electrode's outer surface. In this instance, the surface charge per unit area can be computed. This ubiquitous arrangement might be referred to as electrodes. The presence of at least two distinct phases and a differential in electric potential at the phase boundary are the general characteristics of designed electrodes. Electric current can flow through the electrode in the interim. Silver/silver chloride Ag/AgCl, copper/copper sulfate $Cu/CuSO_4$, mercury/mercurous sulfate Hg/Hg_2SO_4, and mercury/mercuric oxide Hg/HgO are among the various kinds of typical reference electrodes. The potential difference in voltage between the electrode surface and any place with potential can be assumed to be zero. These potentials include electric and open-circuit potentials. Stated otherwise, it is equivalent to the effort that the electric field does to transfer a contacting charge between two phase borders. Since zero potential is assigned to an infinite point, the potential is frequently expressed using the symbol φ. A two-electrode setup is commonly used in electrochemistry to determine an electrode's potential in relation to any reference electrode. At all temperatures, the reference electrode's potential is normally rather stable. One redox electrode that uses the thermodynamics scale of oxidation–reduction potential is the standard hydrogen electrode (SHE). As was covered in the preceding section, the absolute electrode potential at 25 °C is found to be approximately 4.44 V [7].

In electrochemistry, the voltage is typically defined as the difference between the usual reference electrode and the measured electrode potential. The measured potential E is thus defined as a result. Electrode potential calculation in both equilibrium and non-equilibrium situations. As an example, the dissociation of unstable electrode surface charge, which causes the macroscopic current flow through the electrode systems, may be involved in the nonequilibrium situations. The electrode potential is known as the equilibrium potential, or E_{eq}, if the thermodynamic process is carried out at the equilibrium state. The measured potential deviates from the equilibrium value when there is current leakage. The external electric field near the surface of a conducting solid with charge q, material permittivity ε, and spherical radius R is equal to $q/4\pi\varepsilon\varepsilon_o R$. Transferring a single charge from infinity to the electrode phase

β with its potential of ψ_β in vacuum is the same as doing this. Consequently, to finish a charge placement, the work of $q\psi_\beta$ is needed. When considering the ion solvation in this case, the potential increase or reduction is caused by the release or gain of electrons outside the electrode surface in the solution system. For the same phase, the ion solvation contribution to the surface potential χ_β is contributed for solvation potential. Thus, the whole inner potential φ_β can be described as follows:

$$\varphi_\beta = \psi_\beta + \chi_\beta \tag{3.13}$$

Moreover, the potential difference between two potential points in two separate phases or media is known as Volta potential. This potential, for example, can be expressed as follows:

$$\Delta \psi_\beta^\sigma = \psi_\sigma - \psi_\beta \tag{3.14}$$

The difference in electron work functions required to shift a single charge in the electric field at the equilibrium distance between the two phases is the quantitative equivalent of the empirically accessible Volta potential [10]. Similarly, the potential difference between two surface sites in the bulk phases of α and γ is known as the Galvani potential. The potential to be different can be attained through:

$$\Delta \varphi_\gamma^\alpha = \varphi_\alpha - \varphi_\gamma \tag{3.15}$$

The Galvani potential cannot be measured directly using two common electrodes; instead, one needs a reference boundary phase. New physicochemical barriers that occur in the limits would be encountered by the latter measurement [11].

3.4 Foundation of Nernst Response

The determination of the Galvani potential at the metal-liquid contact where the metal's ions are present. Following the ions' thermodynamic equilibrium in the liquid and the metal crystal's oxidation and reduction reactions $M \leftrightarrow M^{z+} + ze^-$. In the equilibrium condition, there is no net ion flux at the metal-solution. For the two metal-solution phases, the chemical potential at the interface should be balanced. It is possible to get,

$$\mu(M)_m + zF\varphi_m = \mu(M^{z+})_s \tag{3.16}$$

The relationship between an ion's electrochemical potential μ_i and its activity μ_i is as follows

$$\mu_i = \mu_i^0 + RT \ln a_i + z_i F \varphi \tag{3.17}$$

The potential difference can be given with the following equation.

$$\Delta\varphi_s^m = \frac{\left(\Delta\mu_i^0\right)_s^m}{zF} + \frac{RT}{zF}\ln\frac{a(M^{z+})}{a(M)} \quad (3.18)$$

The Nernst equation is the one mentioned above. The first term of the potential difference, or standard potential, varies with a constant E^o for each individual reference electrode. Therefore, the general electrode potential results in [12, 13],

$$E = E^o + \frac{RT}{zF}\ln\frac{a(M^{z+})}{a(M)} \quad (3.19)$$

Reconsidering the reversible two half-reactions from the previous section,

$$iA + jB + ne^- \rightleftarrows sC + tD, \quad (3.20)$$

where i, j, s, and t indicate the number of moles of each species, while A, B, C, and D stand for the participating species, such as atoms or ions. Thus, one way to characterize the Nernst expression is as

$$E = E^o + \frac{RT}{nF}\ln\frac{[A]^i[B]^j}{[C]^s[D]^t} \quad (3.21)$$

A general equation that characterizes the entire process in a galvanic cell can be written out to provide a more comprehensive form of derivation.

$$\alpha_1 a_1 + \alpha_2 a_2 + \cdots + \alpha_i a_i \leftrightarrow \alpha_{-1} a_{-1} + \alpha_{-2} a_{-2} + \cdots + \alpha_{-i} a_{-i}, \quad (3.22)$$

where the initial reagent and reactant product activities are denoted by a_i and a_{-i}, respectively. The coefficients α_i and α_{-i} define their stoichiometric factors. The electrode potential then terminates with

$$E = E^o + \frac{RT}{nF}\ln\left(\frac{\Pi\alpha_i a_i}{\Pi\alpha_{-i} a_{-i}}\right) \quad (3.23)$$

It is possible to calculate the electrochemical process's equilibrium constants, K, when the open-circuit voltage is zero. Consequently,

$$E^o - \frac{RT}{nF}\ln\left(\frac{\Pi\alpha_i a_i}{\Pi\alpha_{-i} a_{-i}}\right) = 0 \quad (3.24)$$

It follows that,

$$E^o = \frac{RT}{nF}\ln K = \frac{2.3RT}{nF}\log K \quad (3.25)$$

3.5 Electrode Classification

The value of $E°$ at 25 °C is approximately $(0.059/n) \log K$. In the meantime, the chemical reaction progresses from left to right if $\Delta E > 0$, and from right to left if $\Delta E < 0$. Table 3.1 lists the standard electrode potentials for the various metallic reactions [14, 15].

3.5 Electrode Classification

i. **Electrodes of the first kind**

An electrode of the first kind is created when a metal bar is submerged in an electrolyte and either ions are directly deposited onto the electrode through the reduction process or the metal dissolves into the solution through the oxidation process. The Nernst equation in this instance of the first-kind electrode potential can be expressed as

$$E = E° + \frac{RT}{zF} \ln a(M^{z+}) \tag{3.26}$$

Take a copper electrode, for example, where $Cu^{2+} + 2e^- \rightleftarrows 2Cu$ happens. The reversible reaction on the electrode may occur via a reduction process, however the electrode potential is dependent upon the copper ion activity in the solution.

$$E_{Cu} = E°_{Cu} - \frac{0.0592}{2} \log \frac{1}{a_{Cu^{2+}}} = E°_{Cu} + \frac{0.0592}{2} \log a_{Cu^{2+}}, \tag{3.27}$$

where $a_{Cu^{2+}}$ is the ion's activity in diluted solution, or about its molar concentration, and E_{cu} is the metal electrode's electrode potential. The p-function of the cation is commonly used to express the electrode potential of the indicator electrode ($pCu^{2+} = -\log a_{Cu^{2+}}$). Using this definition of pCu^{2+} as a substitute for any metals or other cations, the general electrode potential (depicted in Fig. 3.4) can be expressed generally as

$$E_M = E°_{M^{z+}} + \frac{0.0592}{z} \log a_{M^{z+}} = E°_{M^{z+}} - \frac{0.0592}{z} pM^{z+} \tag{3.28}$$

For several reasons, first-kind electrode systems are not frequently utilized in electrochemical analyses. Metallic indicator electrodes react to additional cations that are more readily reduced in addition to their own, making them less selective. For instance, because the electrode potential also depends on the concentration of silver, a copper electrode cannot be used to measure cuprite ions in the presence of silver ions. Furthermore, because they disintegrate in the presence of acids, several metal electrodes, including zinc and cadmium, can only be employed in neutral or basic solutions. Some metals can only be employed when analyte solutions are needed to eliminate oxygen because they oxidize so quickly.

Table 3.1 Standard electrode potentials

Half reaction	E^o (V)	Half reaction	E^o (V)
$KrF_2 + 2e^- \rightarrow Kr + 2F^-$	3.27	$2H^+ + 2e^- \rightarrow H_2$	0.00
$Pr^{4+} + e^- \rightarrow Pr^{3+}$	3.20	$Fe^{3+} + 3e^- \rightarrow Fe$	−0.04
$Tb^{4+} + e^- \rightarrow Tb^{3+}$	3.10	$Pb^{3+} + 2e^- \rightarrow Pb$	−0.13
$F_2 + 2H^+ + 2e^- \rightarrow 2HF$	3.07	$Sn^{3+} + 2e^- \rightarrow Sn$	−0.14
$Cm^{4+} + e^- \rightarrow Cm^{3+}$	3.00	$Ni^{3+} + 2e^- \rightarrow Ni$	−0.23
$F_2 + 2e^- \rightarrow 2F^-$	2.87	$V^{3+} + e^- \rightarrow V^{2+}$	−0.26
$O_3 + 2H^+ + 2e^- \rightarrow O_2 + H_2O$	2.07	$Co^{2+} + 2e^- \rightarrow Co$	−0.28
$H_2O_2 + 2H^+ + 2e^- \rightarrow 2H_2O$	1.78	$In^{3+} + 3e^- \rightarrow In$	−0.34
$Au^+ + e^- \rightarrow Au$	1.69	$PbSO_4 + 4H^+ + 2e^- \rightarrow Mn^{2+} + 2H_2O$	−0.36
$Pb^{4+} + 2e^- \rightarrow Pb^{2+}$	1.67	$Cd^{2+} + 2e^- \rightarrow Cd$	−0.40
$Ce^{4+} + e^- \rightarrow Ce^{3+}$	1.61	$Cr^{2+} + e^- \rightarrow Cr^{2+}$	−0.41
$MnO_4^- + 8H^+ + 5e^- \rightarrow Mn^{2+} + 4H_2O$	1.51	$Fe^{2+} + 2e^- \rightarrow Fe$	−0.44
$Au^{3+} + 3e^- \rightarrow Au$	1.40	$U^{4+} + e^- \rightarrow U^{3+}$	−0.61
$Cl_2 + 2e^- \rightarrow 2Cl^-$	1.36	$FeCO_3 + 2e^- \rightarrow Fe + CO_3^{2-}$	−0.75
$O_2 + 4H^+ + 4e^- \rightarrow 2H_2O$	1.23	$Zn^{3+} + 2e^- \rightarrow Zn$	−0.76
$MnO_2 + 4H^+ + 2e^- \rightarrow Mn^{2+} + 2H_2O$	1.21	$2H_2O + 2e^- \rightarrow H_2 + 2OH^-$	−0.83
$Pt^{2+} + 2e^- \rightarrow Pt$	1.20	$Cr^{2+} + 2e^- \rightarrow Cr$	−0.91
$Br_2 + 2e^- \rightarrow 2Br^-$	1.09	$Mn^{2+} + 2e^- \rightarrow Mn$	−1.18
$2Hg^{2+} + 2e^- \rightarrow Hg_2^{2+}$	0.92	$V^{2+} + 2e^- \rightarrow V$	−1.19
$Pd^{2+} + 2e^- \rightarrow Pd$	0.915	$ZnS + 2e^- \rightarrow Zn + S^{2-}$	−1.44
$ClO^- + H_2O + 2e^- \rightarrow Cl^- + 2OH^-$	0.89	$Al^{3+} + 3e^- \rightarrow Al$	−1.66
$Ag^+ + e^- \rightarrow Ag$	0.80	$Mg^{2+} + 2e^- \rightarrow Mg$	−2.36
$Hg_2^{2+} + 2e^- \rightarrow 2Hg$	0.79	$Na^{2+} + e^- \rightarrow Na$	−2.71
$Fe^{3+} + e^- \rightarrow Fe^{2+}$	0.77	$K^+ + e^- \rightarrow K$	−2.92
$MnO_4^- + 2H_2O + 3e^- \rightarrow MnO_2 + 4OH^-$	0.60	$Li^+ + e^- \rightarrow Li$	−3.05
$I_2 + 2e^- \rightarrow 2I^-$	0.54	$3N_2 + 2H^+ + 2e^- \rightarrow 2HN_3$	−3.09
$O_2 + 2H_2O + 4e^- \rightarrow 4OH^-$	0.40	$Pr^{3+} + e^- \rightarrow Pr^{2+}$	−3.10
$Cu^{2+} + 2e^- \rightarrow Cu$	0.34	$Th^{4+} + e^- \rightarrow Th^{3+}$	−3.60
$AgCl + e^- \rightarrow Ag + Cl^-$	0.22	$Ca^+ + e^- \rightarrow Ca$	−3.80
$NO_3^- + H_2O + 2e^- \rightarrow NO_2^- + 2OH^-$	0.01	$Sr^+ + e^- \rightarrow Sr$	−4.10

3.5 Electrode Classification

Fig. 3.4 Theoretical electrode potential obtained for different log multivalent cationic concentrations

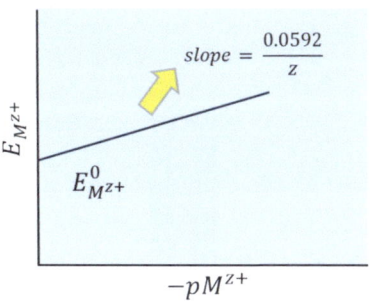

ii. Electrodes of the second kind

Most electrodes used in sensors are of the second to fifth kinds, which differ from the first kind because the metal is coated in insoluble salt or porous polymeric layers. For example, the Ag/AgCl metal can be submerged in an anionic solution containing chlorine ions. At the metal surface, the reaction is $AgCl + e^- \rightarrow Ag + Cl^-$. Although the definition of its solubility product is $K_{sp}(AgCl) = a(Ag^+)a(Cl^-)$, the electrode potential results in

$$E = E^\circ + \frac{RT}{zF} \ln K_{sp} - \frac{RT}{zF} \ln a(Cl^-) = E^\circ_{Ag/AgCl} - \frac{RT}{zF} \ln a(Cl^-) \quad (3.29)$$

The saturated calomel electrode (SCE) given in Fig. 3.5a is another illustration of a second sort electrode. The basis for it is the interaction of mercury(I) chloride with elemental mercury. The notation for its cell is $Cl^-(xM)|Hg_2Cl_2(s)|Hg(l)|Pt$, where x denotes the potassium chloride's molar concentration in the tube.

$$Hg_2Cl_2(s) + 2e^- \rightarrow 2Hg(l) + 2Cl^-(aq) \quad (3.30)$$

iii. Electrodes of the third kind

Redox electrodes are another name for the third type of electrodes. It is an electrode that electrically transfers metal into a solution phase to carry out a reaction. Two ions with distinct electrochemical substances in the electrode liquid are the electrode's active materials. Pt and C are examples of inert conductors that should be utilized as redox electrodes. One type of electrode that is frequently utilized in the field of biological medicine is the quinone hydroquinone electrode. It may be used to measure the pH of biological liquids and is sensitive to the redox reaction of proteins and amino acids. In the solution undertest, the redox electrode exhibits remarkable chemical stability. It can also be applied to electrochemical research to look at the kinetics of electrochemical reactions at the interfacial layer. The Nernst equation describes the relationship between an electrode potential and the composition of the redox system.

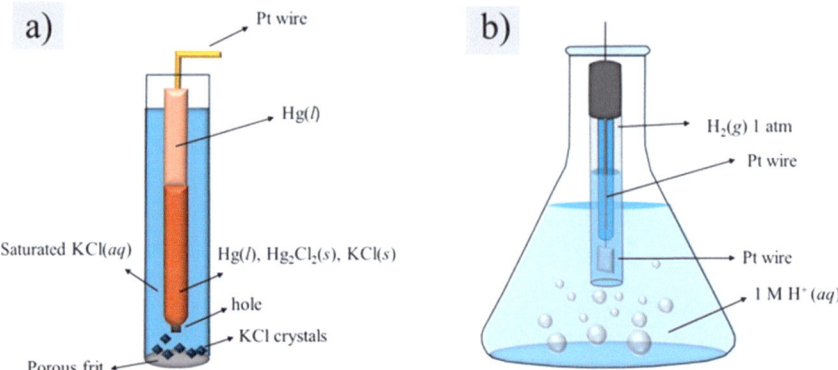

Fig. 3.5 a A typical calomel electrode diagram. **b** A typical standard hydrogen electrode (SHE) diagram

$$E = E^\circ + \frac{RT}{zF} \ln\left(\frac{a_{\text{ox}}}{a_{\text{red}}}\right) \qquad (3.31)$$

Redox electrodes can be divided into two groups in general. Metals (such platinum and gold) come first, but other materials like titanium and indium can also be utilized for certain applications. The semiconductors come in second, with graphite, diamond, and SnO_2 being the most often used types. The redox system includes, for example, typical hydrogen electrodes that use platinum metal. The pressure of the gas determines its potential. One atm is the typical gas pressure. In addition to acting as a catalyst for the electrode reaction and an electron reservoir, the metal in the gas electrode creates electrical contact between the phases. Typically, platinum coated with finely divided platinum black is used for this function. Reversibility of the electrode regarding either the cation or the anion is possible. The reaction $2H_3O^+ + 2e^- \rightarrow H_2 + 2H_2O$ takes place in this electrode. One way to express the Nernst equation is as

$$E = E^\circ + \left(\frac{RT}{2F}\right) \ln \frac{2a(H^+)}{p(H_2)}, \qquad (3.32)$$

where p represents the hydrogen gas's partial pressure. One typical reference electrode is the standard hydrogen electrode (SHE) as exhibited in Fig. 3.5b. This gas is continually passed or bubbled at a constant pressure close to the electrode to achieve a constant amount of adsorbed gas on the electrode surface. In this instance, the hydrogen pressure and ion activity determine the electrode potential.

iv. **Electrodes of the fourth kind**

The membrane electrode is one of the five types of electrodes here. Typically, these electrodes consist of an internal reference electrode, an internal charge liquid, and ion-selective response films as given in Fig. 3.6a. Gas sensitive electrodes, liquid film electrodes, ion exchange membrane electrodes, glass film electrodes and various solid film electrodes are available. The primary feature of this mechanism is the absence of electron transport [17], varying ions have varying permeabilities via membranes. They are separated into two classes based on the permeability mechanism: ion-exchange and porous membranes. Because of their diffusivities, various species have varying levels of permeability through porous membranes. A membrane that allows tiny ions to flow through their holes while blocking larger ones is an example of this type of membrane. Such a membrane that permits hydrogen ions to pass through while obstructing nitrate ions can be created. As a result, the system's charges will be distributed differently [18]. In this instance, ions are redistributed between the solution and the membrane until the system reaches equilibrium and potential jumps form at the membrane electrode and electrolyte boundaries. The most practical way to test pH has been to measure the potential that emerges across a thin glass membrane that divides two solutions with varying quantities of hydrogen ions for almost a century. The measurement is based on a phenomenon that was originally observed in 1906 and has since been thoroughly investigated by other researchers [19]. Consequently, a substantial amount of knowledge exists regarding the sensitivity and selectivity of glass membranes toward hydrogen ions. Moreover, this knowledge has prompted the creation of different kinds of membranes that react differently to a wide range of different ions. Because the data acquired from membrane electrodes are typically displayed as p-functions, such as pH, pCa, or pNO_3, membrane electrodes are sometimes referred to as p-ion or ion selective electrodes [20]. Furthermore, it is crucial to remember that membrane electrodes differ from metal electrodes in both design and principle. To demonstrate these variations, the glass electrode is used for pH measurements. The effects of different glass compositions on the sensitivity of membranes to protons and other cations have been extensively studied, and many formulations are currently employed in the electrode manufacturing process. Approximately 22% Na_2O, 6% CaO, and 72% SiO_2 make up Corning 015 glass, which has been frequently utilized for membranes [21]. Up to a pH of around 12, this glass's membranes show exceptional sensitivity to hydrogen ions. Nevertheless, the glass starts to react to sodium and other singly charged cations at higher pH levels. There are already alternative glass compositions in use that substitute barium and lithium ions for sodium and calcium ions to varying degrees. These membranes are longer-lasting and more selective.

v. **Electrodes of the fifth kind**

Another name for the fifth type of electrodes includes implanted and electrochemical microfluidic electrodes as shown in an example of Fig. 3.6b. This electrode will experience an intercalation reaction, as the name implies. Intercalation reaction is the current reaction occurring on the electrodes of lithium-ion batteries, which are widely used in computers, cell phones, and other electronic devices. These electrodes are the most often used type of electrode in modern electronic items because they can raise the battery's capacity and enhance its conductivity [22].

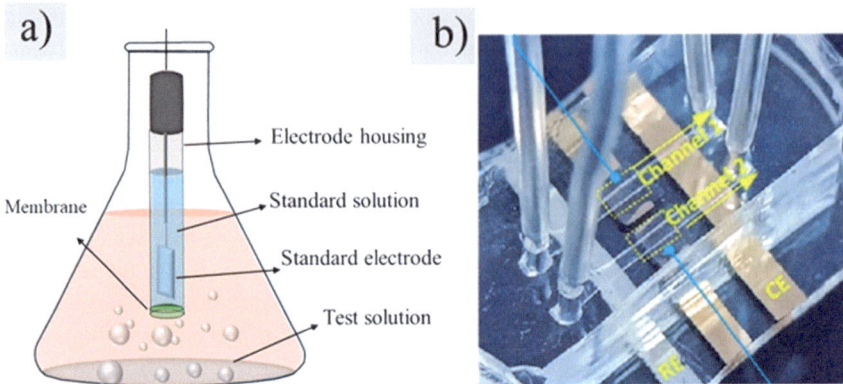

Fig. 3.6 a A typical membrane electrode diagram. **b** Electrochemical microfluidic electrodes of the fifth kind. Adapted with a permission from Ref. [16]. Copyright 2024 American Chemical Society

3.6 Electrochemical Cell

Two electrically conducting electrodes submerged in or in contact with an electrolyte make up an electrochemical cell. The electrodes conduct electrons because they are electronic conductors, also known as semiconductors. Ions, or charged atoms or molecules, are required to transfer the charge between electrodes because electrons are unable to transmit charge in solutions or salts. Ions in the electrolyte participate in heterogeneous processes on electrode surfaces, which cause electrons to move to or from the conductive electrodes. Figure 3.7 illustrates the electrochemical cell. It shows two electrodes submerged in a solution with charge carries positive and negative ions in the electrolyte. It is demonstrated how the electrons flow through the external circuit, entering the electrode where reduction happens and exiting the electrode where oxidation occurs. The galvanic cell, which produces power and is indicated by resistance, the electrolytic cell, which affects changes in the solution and produces chemicals, and the electroanalytical cell, which measures current and voltage, are all indicated by the external driving voltage. The fact that charge transfer takes place at the interface between the solid electrode and liquid electrolyte is one of the special features of electrochemical reactions. An electrochemical cell is a device that either uses electricity to drive a nonspontaneous process or generates power from a spontaneous oxidation–reduction reaction. Because it applies to the concepts of electrochemistry and is the smallest operational unit of an electrochemical system, an electrochemical cell gets its name. Electrons are moved from one species to another during redox reactions. When a reaction occurs spontaneously, energy is liberated and can be put to good use. If the reaction is not spontaneous, the system needs to operate, and energy is used [23, 24]. An electrochemical cell is a device that either uses electricity to drive a nonspontaneous process or generates power from a spontaneous oxidation–reduction reaction. Because it applies to the concepts of electrochemistry and is the smallest operational unit of an electrochemical system,

3.7 Electrode Reaction

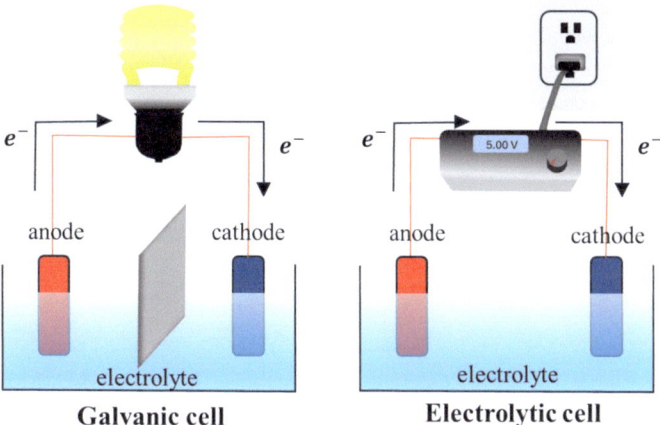

Fig. 3.7 Diagrams displaying galvanic and electrolytic cells, two types of electrochemical cells

an electrochemical cell gets its name. The study of electricity and its relationship to chemical reactions is known as electrochemistry. In electrochemistry, a process called an oxidation–reduction reaction, or redox reaction, can produce energy by moving electrons from one element to another. A chemical process that occurs spontaneously powers galvanic cells. This indicates that electrons will move freely between the electrochemical cell's two sides. Devices like batteries and fuel cells are built on top of these cells. However, there are other types of electrochemical cells as well. A nonspontaneous chemical reaction occurs in an electrolytic cell. Electrons that are stuck in a nonspontaneous reaction do not want to move. Therefore, to induce electrons to flow, an electrolytic cell needs an external electrical energy source. A potential difference between the metal electrodes results from nonspontaneous driving the redox process. Rechargeable batteries function as electrolytic cells when they are being recharged, but they function as galvanic cells when they are draining electrical energy. The cathode is the electrode where electrons that are reactants in the reaction and enter from the external circuit. The cathode is the site of reduction. The anode is the electrode from which electrons depart. Electrons are the result of the process when oxidation occurs on the anode [25, 26].

3.7 Electrode Reaction

The reaction that happens from reactants on the left side of the arrow to products on the right side is referred to as a typical chemical reaction. According to the sign convention of the International Union of Pure and Applied Chemistry (IUPAC), the cell reaction is also thought to proceed in a specific direction when we examine an electrochemical cell and its potential. The plus right rule is the name of the convention for cells. According to the rule, the positive contact of the voltmeter should always

Fig. 3.8 Shift in the cell potential following the passage of current in the high-resistance voltmeter with an open-circuit potential of 0.4324 V until equilibrium is established

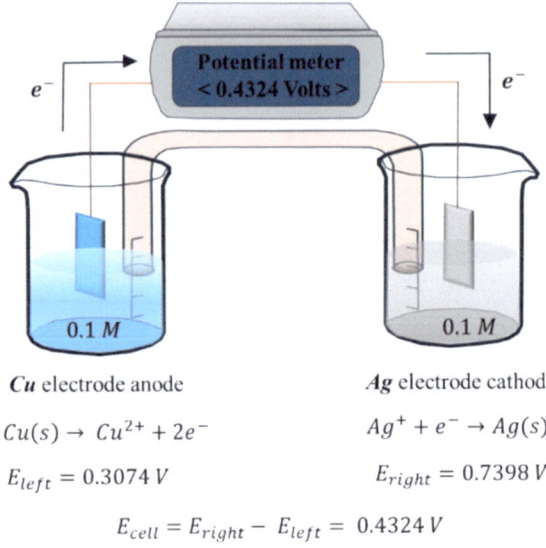

Cu electrode anode

$Cu(s) \rightarrow Cu^{2+} + 2e^-$

$E_{left} = 0.3074\ V$

Ag electrode cathode

$Ag^+ + e^- \rightarrow Ag(s)$

$E_{right} = 0.7398\ V$

$E_{cell} = E_{right} - E_{left} = 0.4324\ V$

be connected to the right-hand electrode (Ag) in the schematic picture of Fig. 3.8, and the ground contact should always be connected to the left-hand electrode (Cu). According to this convention, the tendency of the cell reaction to occur spontaneously in the left-to-right direction is measured by the value of E_{cell}. The standard galvanic cell in this instance can be expressed as

$$Cu|Cu^{2+}(0.1\ M)||Ag^+(0.1\ M)|Ag \quad (3.33)$$

In this situation, Cu metal is oxidized to Cu^{2+} in the left electrode of the reaction, whereas Ag^+ is reduced to Ag metal in the right electrode. This is the overall process direction. Nonetheless, the response might be regarded as

$$Cu(s) + 2Ag^+ \rightleftarrows Cu^{2+} + 2Ag(s) \quad (3.34)$$

The difference between two half-cell or single-electrode potentials, one connected to the half-reaction at the right-hand electrode (E_{right}) and the other to the half-reaction at the left-hand electrode (E_{left}), represents the potential of a cell, such as that displayed in a conventional cell. If there is no liquid junction or the liquid-junction potential is minimal, we can write the cell potential E_{cell} as per the IUPAC sign convention.

$$E_{cell} = E_{right} - E_{left} \quad (3.35)$$

3.8 Electrocatalysis

Heterogeneous catalysis that increases the rate of a reaction at the electrode–electrolyte contact is known as electrocatalysis. The electrode surface or the electrode surface coated in catalytic species, such as metallic particles and enzymes, acts as the catalyst in this kind of catalysis. The catalyst aids in the transport of electrons from the electrode to the reactants or helps the adsorption intermediate that is created during the reaction undergo a chemical change. As given in Fig. 3.9a, R_1 and R_2 are the reactants required to overcome the activation energy to form product P. Chemical processes proceed more quickly when a catalyst C is present in Fig. 3.9b because it offers a different, lower-activation-energy reaction pathway than when there is no catalyst. The energy and reaction pathway changes that occur during a catalyzed reaction are depicted in the figure. Since catalysts have the same influence on the rate constant of the forward k_f and reverse k_r processes, they have no effect on the chemical equilibrium and do not alter the equilibrium constant $K = \frac{k_f}{k_r}$.

Consequently, the catalysts have no effects on the reaction's standard Gibbs free energy, ΔG [27–29]. Many catalytic materials, such as Pt, Pd, Ru, Rh, or metal oxides, are utilized in both types of catalysis. The catalyst's surface area and chemical makeup, for example, affect the rate of reaction in both situations. Additionally, in both situations, the catalyst quickens the slow step or modifies the reaction route to speed up the process. What sets chemical catalysis apart from electrocatalysis? The

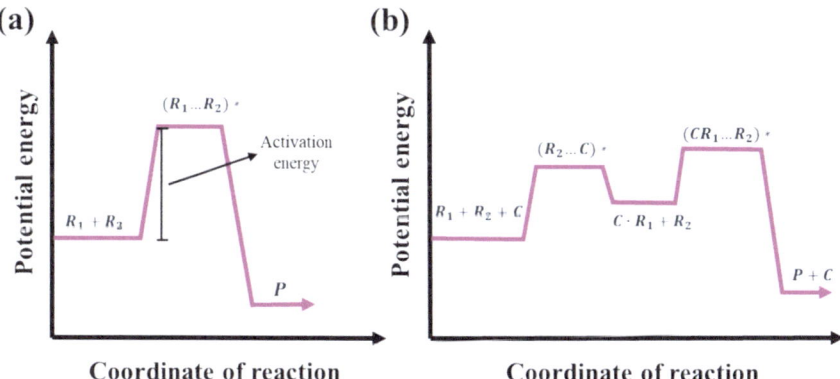

Fig. 3.9 Reactants (R_1 and R_2), product (P), catalyst (C), and changes in energy and reaction routes between non-catalyzed (**a**) and catalyzed (**b**) chemical reactions are depicted in the schemes

primary distinction is the electrocatalytic reaction rate's reliance on electrode potential, or more precisely, the potential difference across the interface, which is adjustable via an external voltage source. The rates v_r of chemical and electrochemical processes as well as the approximate rates of first-order catalytic and electrocatalytic reactions are presented in the following equations [30].

$$\text{Chemical catalysis:} \quad R_1 \xrightarrow{kf_1} R_2; \quad v_r = C_{R_1} k_{f_1}; \quad v_r = AC_{R_1} e^{\frac{-\Delta G}{RT}} \tag{3.36}$$

$$\text{Electrocatalysis:} \quad R_1^{n+} + ne^- \xrightarrow{kf_2} R_2; \quad v_r = nFC_{R_1} k_{f_2}$$

$$v_r = \frac{j}{nF} = AC_{R_1} e^{\frac{-\Delta G}{RT}} e^{\frac{-\alpha nFE}{RT}} \tag{3.37}$$

The parameters are as follows: A is the preexponential factor, E is the electrode potential, j is the current density, α is the transfer coefficient, and ΔG is the Gibbs free energy of activation during catalysis process. As can be seen from the previous calculations, even by several orders of magnitude, tiny changes in electrode potential result in a massive increase in reaction rate. Furthermore, metal entities with a highly limited size distribution can be created using innovative methods for materials synthesis. These accomplishments may now be seen not only through electron microscopy but also by analysis of the coordination environment of those metal species during reaction conditions using X-ray absorption spectroscopy (XAS). Although a great deal of knowledge has been gained about metal catalysts and a significant amount of experimental work has been conducted, we still lack a single, comprehensive theory that can describe and forecast the behavior of various metal catalysts with varying particle sizes for various processes.

References

1. F.C. Strong, Faraday's laws in one equation. J. Chem. Educ. **38**(2), 98 (1961)
2. G. Wilcox, D. Gabe, Faraday's laws of electrolysis. Trans. IMF **70**(2), 93–94 (1992)
3. A.P. Brown, F.C. Anson, Electron transfer kinetics with both reactant and product attached to the electrode surface. J. Electroanal. Chem. Interfacial Electrochem. **92**(2), 133–145 (1978)
4. S. Hernández-Rizo, E. Larios-Durán, M. Bárcena-Soto, Frequency response of Gibbs free energy and enthalpy changes of electrochemical systems analyzed as thermometric transfer functions. J. Solid State Electrochem. **27**(11), 3177–3188 (2023)
5. V. Tripkovic et al., Standard hydrogen electrode and potential of zero charge in density functional calculations. Phys. Rev. B Condens. Matter Mater. Phys. **84**(11), 115452 (2011)
6. X. Jin, G.G. Botte, Electrochemical technique to measure Fe(II) and Fe(III) concentrations simultaneously. J. Appl. Electrochem. **39**, 1709–1717 (2009)
7. H. Reiss, A. Heller, The absolute potential of the standard hydrogen electrode: a new estimate. J. Phys. Chem. **89**(20), 4207–4213 (1985)
8. E.J. Dickinson, A.J. Wain, The Butler–Volmer equation in electrochemical theory: origins, value, and practical application. J. Electroanal. Chem. **872**, 114145 (2020)

9. M.R. Shaner, K.T. Fountaine, H.-J. Lewerenz, Current-voltage characteristics of coupled photodiode-electrocatalyst devices. Appl. Phys. Lett. **103**(14) (2013)
10. S. Ross, The story of the Volta potential, in *Nineteenth-Century Attitudes: Men of Science* (Springer, 1978), pp. 40–83
11. P. Rüetschi, The Galvani electrode potential. J. Electrochem. Soc. **104**(3), 176 (1957)
12. F.J. Vidal-Iglesias et al., Understanding the Nernst equation and other electrochemical concepts: an easy experimental approach for students. J. Chem. Educ. **89**(7), 936–939 (2012)
13. A.-S. Feiner, A. McEvoy, The Nernst equation. J. Chem. Educ. **71**(6), 493 (1994)
14. G. Milazzo, S. Caroli, R.D. Braun, Tables of standard electrode potentials. J. Electrochem. Soc. **125**(6), 261C (1978)
15. S.G. Bratsch, Standard electrode potentials and temperature coefficients in water at 298.15 K. J. Phys. Chem. Ref. Data **18**(1), 1–21 (1989)
16. N. Singh et al., Electrochemical and plasmonic detection of myocardial infarction using microfluidic biochip incorporated with mesoporous nanoscaffolds. ACS Appl. Mater. Interfaces (2024)
17. M. Huser et al., Membrane technology and dynamic response of ion-selective liquid-membrane electrodes. Anal. Chem. **63**(14), 1380–1386 (1991)
18. W.E. Morf, *The Principles of Ion-Selective Electrodes and of Membrane Transport* (Elsevier, 2012)
19. J.R. Allen, pH electrodes, ion-selective electrodes, and oxygen sensors: electrochemical sensors used in the medical field. Lab. Med. **34**(7), 544–547 (2003)
20. G.H. Fricke, Ion-selective electrodes. Anal. Chem. **52**(5), 259–275 (1980)
21. H. Dunken, R.H. Doremus, Short time reactions of a Na_2O-CaO-SiO_2 glass with water and salt solutions. J. Non-Cryst. Solids **92**(1), 61–72 (1987)
22. R. Jalili et al., Implantable electrodes. Curr. Opin. Electrochem. **3**(1), 68–74 (2017)
23. H. Oberacher et al., Mass spectrometric methods for monitoring redox processes in electrochemical cells. Mass Spectrom. Rev. **34**(1), 64–92 (2015)
24. T. Ikeda, K. Kano, An electrochemical approach to the studies of biological redox reactions and their applications to biosensors, bioreactors, and biofuel cells. J. Biosci. Bioeng. **92**(1), 9–18 (2001)
25. Y. Zhang et al., Materials design and fundamental understanding of tellurium-based electrochemistry for rechargeable batteries. Energy Storage Mater. **40**, 166–188 (2021)
26. J.R. Miller, Perspective on electrochemical capacitor energy storage. Appl. Surf. Sci. **460**, 3–7 (2018)
27. S. Zhao, Y. Yang, Z. Tang, Insight into structural evolution, active sites, and stability of heterogeneous electrocatalysts. Angew. Chem. Int. Ed. **61**(11), e202110186 (2022)
28. S. Hammes-Schiffer, G. Galli, Integration of theory and experiment in the modelling of heterogeneous electrocatalysis. Nat. Energy **6**(7), 700–705 (2021)
29. Z. Levell et al., Emerging atomistic modeling methods for heterogeneous electrocatalysis. Chem. Rev. (2024)
30. Y. Xia et al., *Introduction: Advanced Materials and Methods for Catalysis and Electrocatalysis by Transition Metals* (ACS Publications, 2021), pp. 563–566

Chapter 4
Tools in Nanotechnology for Sensors

Abstract Exploring features at the nanoscale or even the atomic scale is a common practice in several nanotechnology-related fields. For example, the field of lab-on-a-chip electronics manufacture is already moving from microelectronics to nanoelectronics. Since transistors with a substantial 3-nm length are now being produced. These transistors are essential components found in practically all electrical devices. Many of them can fit inside a single laptop, smartphone, or even lab-on-a-chip, suggesting that even small variations in size can influence these gadgets. In the meantime, for many years, a variety of sensing, imaging, and detection applications have made use of X-rays and spectroscopic techniques. At the nano and even atomic scales, these well-established instruments are capable of characterizing, identifying chemical structure, and conducting morphological tests on the miniature devices and nanostructures. The method for imaging surfaces at the atomic level utilizing the idea of quantum tunneling would be covered in detail in the first chapter.

4.1 Scanning Tunneling Microscope (STM)

Gerd Binnig and Heinrich Rohrer created the scanning tunneling microscope (STM) for the first time in 1981 at the IBM laboratories in Zurich and later received the Nobel Prize in physics 1986 for this invention. The main concept is rather straightforward, much like many famous inventions. Using an incredibly sharp conducting tip, STM detects surface characteristics with a 10 pm depth resolution, allowing it to differentiate structures smaller than 1 Å. The STM microscope as given in Fig. 4.1a is being advanced by a quantum mechanical phenomenon known as tunneling. When electrons pass through a barrier that they shouldn't be able to pass through according to classical theory, a tunneling current happens. In classical words, one cannot move over a barrier if one does not have the energy to do so. But in the domain of quantum mechanics, electrons are like waves. These waves taper off quickly rather than suddenly stopping at a wall or other barrier. The probability function may pass through the barrier and enter the following zone if it is sufficiently thin. Given enough electrons, some will really pass through the barrier and appear on the other

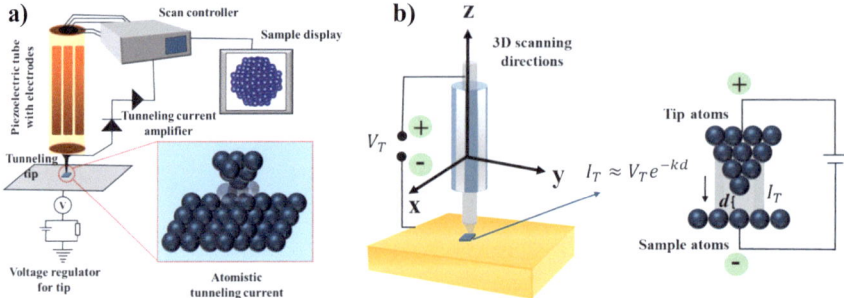

Fig. 4.1 a Principal operation of scanning tunneling microscope. **b** Tunneling voltage with scanning direction

side despite the tiny chance that any electrons are on the opposite side. This type of movement of an electron through the barrier is known as tunneling, as was covered in the previous chapter [1, 2].

A single atom or nanoclusters should ideally protrude from the tip of every scanning probe microscope, which is essentially a sharp tip that has been etched or processed to have a radius of a few nanometers as shown in Fig. 4.1b. The tip is positioned close to the sample using a coarse positioning mechanism that is often verified visually. With piezoelectric scanning tubes, the tip location in relation to the sample surface may be precisely controlled. At close range, piezoelectric scanning tubes, whose approaching separation may be adjusted by a control voltage and provide fine control of the tip position regarding the sample surface. The scanner is progressively extended until the tip begins to receive the tunneling current after a bias voltage is placed between the sample and the tip. The tip-to-sample separation is then maintained within the range of 3–10 Å in which the tip encounters attractive forces. This region lies slightly above the 3 Å threshold where repulsive interaction occurs at the tip. The sub-nanoampere tunneling current is boosted as close to the scanner as feasible. $I_T = V_T e^{-kd}$ is an estimate for the tunneling current I_T given a tip/sample distance of d and an applied tunneling voltage of V_T. Here, k is a metallic constant in this instance [3, 4].

After tunneling is established, experimental calibration is considered when varying the sample bias and tip position with relation to the sample are needed. The tunneling current varies as the tip moves over the surface in a discrete x–y matrix due to variations in the discrete electronic states and surface height. There are two methods used to create digital images of the surface, including the constant-current and constant-height modes. In this regard, the constant-current mode modulates voltage controlling the tip height and records the traced surface height while the tunneling current is maintained at the preset level. To keep the tunneling current constant, a feedback loop continuously measures it and modifies the tip movement. The computer records these modifications, which the STM program can display a topographical structure with calibrated tunneling current. During this scanning, a steady tunneling current of a few nano amperes are maintained. It is most frequently

4.2 Atomic Force Microscopy (AFM)

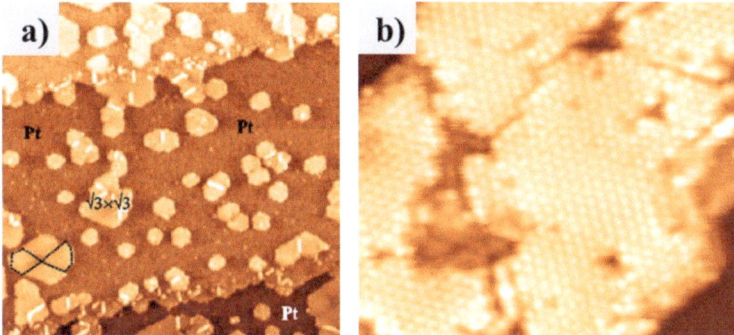

Fig. 4.2 Atomic resolution STM topographic images of **a** Pt(111) and **b** CrO$_x$ phase on Pt(111). Reproduced with a permission from Ref. [5]. Copyright 2024 CC-BY 4.0 American Chemical Society

used in STM. Examples of STM Pt(111) and chromium oxide phase on Pt(111) images are given in Fig. 4.2 [6, 7]. Meanwhile, the constant-height mode traced the tunneling current which is exponentially dependent on separation distance. The z-scanner voltage is fixed in this method. The image is formed by a variation in tunneling current. Faster imaging is possible with this method, although it is limited to flat samples. When relying on this mode, the intense rough surface may put the tip at risk of crashing. Further information on the electrical structure and density of states of the sample can be obtained by scanning tunneling spectroscopy (STS) through sweeping the bias voltage with a little amount of alternating current (AC) modulation. This STS scan can perform very local measures and detects changes in current at a given area. Modern scanning tunneling microscopes can capture images at rapid frames per second. These photos can be used to create videos that follow adsorption and reactions on the surface or demonstrate surface diffusion [8–10]. The attachment of the tip on a piezoelectric plate allows for very delicate motions of the tip. With the help of the coating metal, which shrinks in response to voltage, precise movements with sub-nanometer accuracy can be achieved. To identify a change in electrical impedance between the outermost tip atom in the probe and the surface atoms in a conducting material, the STM uses the quantum mechanical tunneling phenomenon [11].

4.2 Atomic Force Microscopy (AFM)

One drawback of STM is that, because tunneling current is the quantity that is measured, it can only be utilized for conducting samples. Hence, STM has inability to capture images from insulating surfaces, Binnig et al. in 1986 developed the atomic force microscope (AFM), which is depicted in Fig. 4.3. Cantilever deflections of about 1 Å can be observed due to an optical lever effect that is long enough. Atomic force microscopy, a type of scanning probe microscopy (SPM), has been reported to have a resolution on the scale of nanometer fractions, exceeding the optical diffraction limit by more than a thousand times. Data is obtained by feeling or touching the surface with a mechanical probe. Piezoelectric components enable small, precise, and controlled movements under electric control, which enables accurate scanning. An atomic force microscope does not use nuclear force, despite its name. A typical commercial cantilever is made on a silicon chip of between 100 and 500 μm in length, 30–50 μm in width, and 0.5–8 μm in thickness shown in Fig. 4.4a. The free-swinging end of a cantilever that extends from a holder is fitted with an AFM probe, which has a pointed tip. This provides a solid foundation for installing the cantilever within the microscope. The cantilever length varies based on the application once the tip is generated at the end. Applied force for nanoparticle and cell interactions might be calculated using [12–14]

$$F_{tip} = kq(z). \tag{4.1}$$

The applied force F, the cantilever spring constant k, and the cantilever deflection distance, $q(z)$ are all represented in this equation. The engineered CeO_2 and Fe_2O_3 nanoparticles were used to attached on AFM tips. The atomic force between the tip and lung epithelia cells adhered on a substrate was examined in biological media. The existence of serum proteins minimized both the detachment force and the tip bonding [15]. For instance, constructed AFM probe as given in Fig. 4.4b was used to measure substrate and elastic modulus of layered films. Compared to the ordinary pyramid (Fig. 4.4c) and reactive ion etching (RIE) (Fig. 4.4d) tips, the AFM rocket tip (Fig. 4.4e) produced a greater side-wall imaging resolution due to its exceptional geometrical design.

A laser is reflected from the cantilever's back into a position-sensitive deflector to measure the deflection of the cantilever. Based on the type of tip motion, the AFM

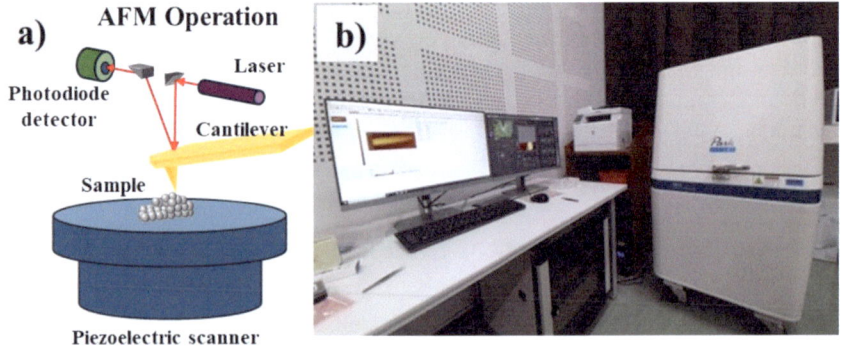

Fig. 4.3 Schematic shows **a** an overall operation of atomic force microscopy (AFM) and **b** an instrumentation photograph of the AFM. Courtesy King Mongkut's University of Technology Thonburi

4.2 Atomic Force Microscopy (AFM)

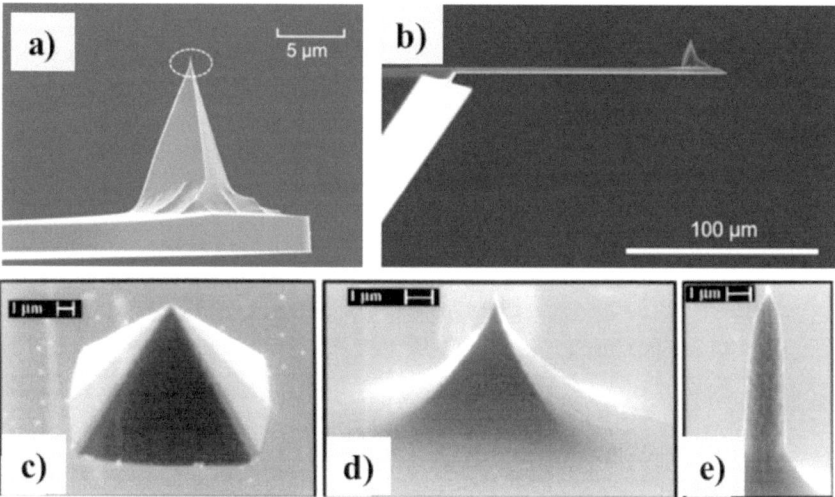

Fig. 4.4 **a** An overall AFM cantilever dimension (model: RFESPA-75). Reproduced with a permission from Ref. [16]. Copyright 2023 CC-BY 4.0 American Chemical Society. **b** A constructed AFM probe used to measure subsurface structure. Reproduced with a permission from Ref. [17]. Copyright 2022 American Chemical Society. SEM images of **c** standard pyramid tip, **d** reactive ion etching (RIE) tip and **e** rocket tip. Reproduced from Ref. [18]. Copyright 2024 American Chemical Society

tip operation can be broadly categorized into three modes including tapping, contact and non-contact modes [19, 20]. The tapping mode is also known as the vibrating or intermittent contact mode. The resonant frequency is defined as

$$\Delta f = \frac{f_0}{2k} \frac{df}{dz} \quad (4.2)$$

Though it oscillates just above the surface, the cantilever oscillates at a much greater amplitude. More oscillation produces a deflection signal large enough for the control circuit, which facilitates topographical feedback control. It produces mediocre AFM results, but it blunts the tip's sharpness faster, which ultimately speeds up the loss of imaging resolution. The contact or static mode is the most straightforward way to retrieve the sample topography. Contact mode often uses cantilever bending for feedback with a somewhat low spring constant to avoid sample damage. Because of the repulsive contact forces between the tip and sample during contact mode imaging, the cantilever bends to accommodate changes in topography. The non-contact or frequency modulation AFM mode preserves the sharpness of the sample surface and tip and yields more accurate results. In this mode, a piezoelectric modulator causes a cantilever to vibrate at a predetermined frequency and low amplitude close to its resonance frequency [21, 22]. When the cantilever is in contact mode, as seen in Fig. 4.5, its deflection is measured while the tip is scanned across the surface

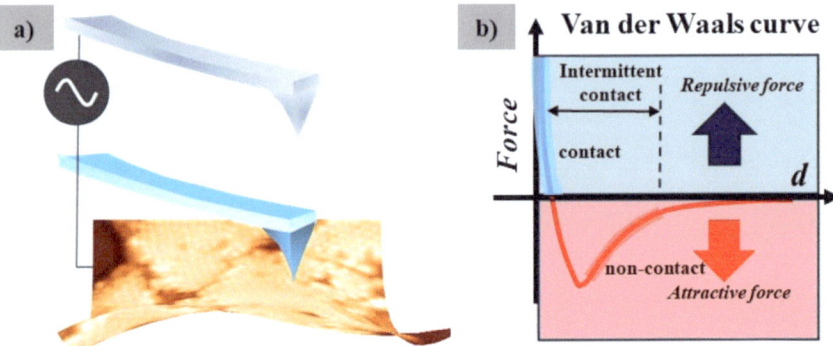

Fig. 4.5 a Contact imaging mode and **b** van der Waals force range

and plotted as a function of x and y to produce an image. The cantilever vibrates in the non-contact mode (bottom) at its resonance frequency, which is typically between 10 and 100 kHz. The frequency shift, which is proportionate to the force gradient, is plotted as a function of x and y to produce the image. A probe is brought near the surface by the AFM. The deflection of a spring, often a cantilever (diving board), is used to measure the force. The distance between the probe tip and the sample is adjusted by sensing forces between them. When using contact mode, the cantilever's deflection is used to assess the surface contours as the tip is "dragged" across the sample. Attractive forces can be quite strong near the sample's surface, which causes the tip to "snap-in" to the surface. Therefore, contact mode AFM is nearly always performed in firm "contact" with the solid surface, that is, at a depth where the total force is repulsive. The cantilever's tip does not contact the surface of the sample. Instead, the cantilever oscillates at its resonant frequency (frequency modulation) or slightly above it (amplitude modulation), with the oscillation's amplitude commonly ranging from a few nanometers (< 10 nm) to a few picometers [23–25].

4.3 Dip-Pen Nanolithography (DPN)

Depending on the surface's wetting characteristics and relative humidity, a dynamic mechanism moves water from the tip to the surface or the other way around. They showed that it was possible to deposit stable water layers by advancing the tip across an appropriate substrate. It is feasible to "write" lines that are about 30 nm wide in a way like a dip-pen operation at the nanoscale if the tip is pre-loaded with a soluble molecule. In this case, the water delivers the molecules onto the surface at the tip position under the correct circumstances. DPN as given in Fig. 4.6a is essentially a materials transport technique in which materials are delivered to a substrate via a meniscus connecting them, using scanning probes (tips) coated with ink. Several variables, including the ink's chemical and physical makeup, surface

4.3 Dip-Pen Nanolithography (DPN)

tension, interfacial wettability, tip structure, ink coverage on the tip, Laplace pressure at the meniscus, ambient humidity, temperature, and operational parameters (such as dwell time, contact force, and lifting speed), can affect the intricate physical mechanism of this transport process. Of these, the ink's material characteristics are crucial, as it can be categorized as either liquid or diffusive. As the name implies, fluid flow dominates the transport of liquid inks, whereas molecular diffusion dominates the transport of diffusive inks. Because they are essential to the DPN process, the two distinct transport process types are outlined here. To select inks that are suitable to the desired DPN result and, if needed, to create new ink materials, the researcher must have a thorough understanding of ink transport. When it comes to alkanethiols, diffuse inks usually form a self-assembled monolayer (SAM) on the surface and are soluble in water. Four processes make up a typical DPN experiment: patterning, solvent evaporation, humidity control, and dip-coating. The ink transport process begins when an ink-coated tip contacts the substrate, forming a condensed water meniscus at the interface. To better understand the development and operation of the water meniscus, several physical models have been developed to examine the transport processes of small molecule inks. The manufacture of arrays of active biological molecules for the ultrasensitive detection of viruses or antigens is one of the primary technological uses of DPN, as shown in Fig. 4.6b.

First, an array of dots inked with a chemical that will bind antibodies is created using the AFM tip. Mercaptohexadecanoic acid (MHA), with the chemical formula $HS(CH_2)_{15}CO_2H$, is a suitable material with high antibody binding. Because DPN is a potent generic approach for synthesizing arrays of a variety of biological components, including DNA, peptides, and proteins, it is obvious that DPN will be a significant tool for medical diagnosis. Over the last two decades, DPN has undergone remarkable development, going from a low-throughput serial writing tool to a massively parallel nanofabrication and chemical synthesis platform as provided in Fig. 4.7. To top it off, DPN has served as the foundation for the creation of patterning methodologies based on cantilever-free tip arrays. Furthermore, scientists have created additional iterations of DPN that rely on outside energy sources for patterning, including heat, mechanical force, electricity, and photons. DPN is a special instrument for nanolithography that has numerous benefits over traditional lithographic techniques, such as high resolution, high registration, and low cost. It may be used to regulate the transport of a wide range of materials from scanning

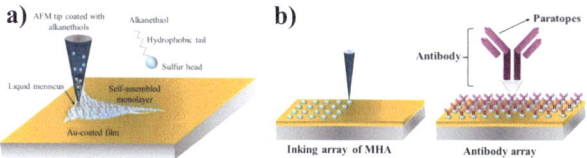

Fig. 4.6 **a** The traditional DPN process involves molecular ink spreading via a water meniscus from a nanoscale tip to the surface. **b** Dip-pen nanolithography with inking array and antibody array printing

probes to surfaces. This means that high-precision leveling operations and ultrafine tip array creation are both required from the standpoint of creating tips. Other polymers, like poly(methyl methacrylate) (PMMA), poly([methyl methacrylate]-co-[butylmethacrylate]) (PMMA-co-PBMA), and poly(3 mercapto propylmethylsiloxane) (PMMS), can also be employed to produce polymer pen lithography (PPL) tip arrays in terms of their material composition. It is not surprising that people frequently believe that any commercial AFM can do DPN experiments because DPN immediately emerged from AFM. Apparently, an AFM is not necessary for DPN, and the AFM does not always have true DPN capabilities [27]. Electron beam (E-beam) lithography and scanning electron microscopy (SEM) have a great resemblance.

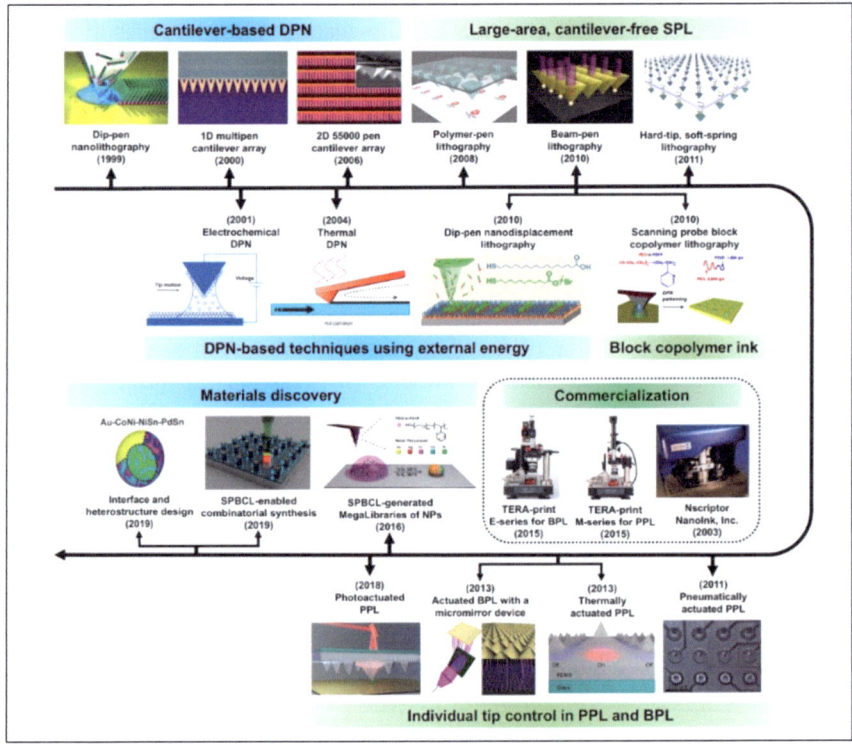

Fig. 4.7 Timeline showing the evolution of DPN instrument. Reproduced with the permission from Ref. [26]. Copyright 2020 American Chemical Society

4.4 Transmission Electron Microscopy (TEM)

Most people would say that a light microscope is a device used to magnify objects that are too small to view with the human eye, and they would most likely be talking about visible-light microscopes (VLMs). Wherever it is instructive, one can draw comparisons between electron and light microscopes due to the widespread familiarity with the VLM idea. The shortest distance that humans can resolve with their eyes is approximately 0.02–0.2 mm at the distance of 15 cm from a human face, or two lines spaced 0.01° apart, depending on how well one's eyesight is working and the amount of light present. The resolving power of the eyes determines the resolution or refined distance. Commonly, VLM microscope is any device that can display pictures or images with details finer than 0.1 mm. The idea feasible to create a microscope that could see detail well below the atomic level was one of the main draws for the early Transmission electron microscopy (TEM) developers, since electrons are smaller than atoms [28, 29]. The method of TEM microscopy and TEM instrumentation depicted in Fig. 4.8 that creates an image by passing an electron beam through a sample. The top-down component of the TEM is the emission source, or cathode, which can be a field emission cannon, a single lanthanum hexaboride crystal, or a tungsten filament. The cannon discharges electrons into the vacuum either by field electron emission or thermionic emission when it is linked to a high voltage source, usually between 100 and 300 kV. There are two ways to manipulate the electron beam: through physical effects. Electromagnets can control the electron beam because of the left-hand rule, which describes how electrons flow while interacting with a magnetic field that has been started in a condensed lens. Electrons may also be deflected through a constant angle by electrostatic forces. Beam shifting is made possible by the production of a shift in the beam path when two opposing deflections are coupled with a little intermediate gap [30].

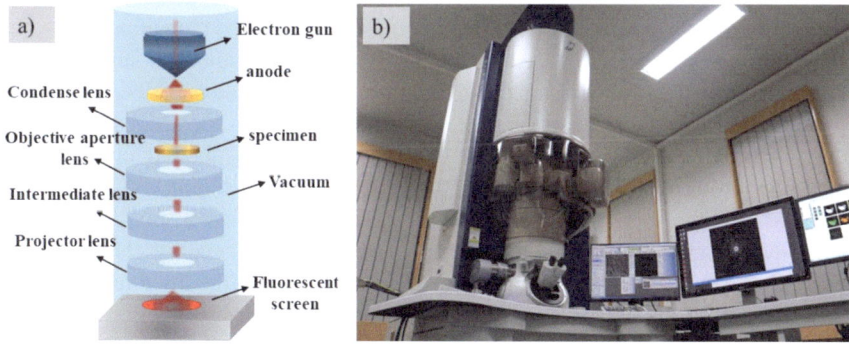

Fig. 4.8 **a** Fundamental operation of TEM and **b** TEM instrumentation (JEOL/JEM-ARM200F). Courtesy Thailand Scientific Equipment Center Network

4.5 X-Ray Photoelectron Spectroscopy (XPS)

High energy X-ray photons are used in X-ray photoelectron spectroscopy (XPS) with a photon energy of 200–2000 eV to excite core electrons in the near-surface region. On the other hand, lower energy photons in the deep UV area are used in ultraviolet photoelectron spectroscopy (UPS) with a photon energy of 10–45 eV to excite valence electrons [31]. Because most of the produced photoelectrons do not escape the solid, XPS and UPS have modest sample depths. Below the surface, the photoelectrons are dispersed and undetectable.

Only those that are 1–10 nm from the outer surface and are amenable to analysis. Photons of fixed energy or monochromatic radiation sources are used in XPS as shown in Fig. 4.9a. When a photon interacts with a solid's outer surface valence levels in UPS, one of these valence electrons is removed, causing ionization. A photoelectron spectrum can be captured by measuring the kinetic energy distribution of the released photoelectrons using any suitable electron energy analyser [32]. XPS is a kind of photoemission spectroscopy where a material is exposed to an X-ray beam to create an electron population spectrum. Chemical states can be inferred by monitoring the kinetic energy and the number of ejected electrons. While there are efforts to produce ambient-pressure XPS, which analyses materials at pressures as low as a few tens of millibar, it necessitates a high vacuum with a residual gas pressure of approximately 100 Pa. Except for hydrogen and helium, all elements may be readily identified by XPS utilizing X-ray sources in a lab [33]. Parts per million (ppm) can be obtained with extended collection times and concentration at the upper surface, but the detection limit is in the parts per thousand range. A wide range of materials are commonly analyzed with XPS instrumentation (Fig. 4.9b), including inorganic compounds, metal alloys, polymers, elements, catalysts, glasses, ceramics, paints, papers, inks, woods, plant parts, makeup, teeth, bones, medical implants, biomaterials, coatings, viscous oils, glues, and ion-modified materials. Less frequently, XPS is used to analyze the hydrated forms of materials like hydrogels and biological samples by freezing them in their hydrated condition in an ultrapure atmosphere and allowing numerous layers of ice to sublime away before analysis to sublime away before analysis [31]. An investigated spectrum of silicon oxide on silicon wafer was created utilizing a wide energy range to illustrate an XPS spectrum as provided from a specimen displayed in Fig. 4.10a. The spectrum displays many peaks for silicon and oxygen. With a modern XPS spectrometer, the silicon oxide spectrum can be obtained in about 10 s. Since silicon oxide is an electrical insulator, specimen charge was managed throughout analysis by shining a low energy electron beam on it. A schematic representation of an XPS instrument and the three most popular ways to transmit the data are displayed in Fig. 4.10b. First, any electron emissions that fall inside a certain energy range and their energy distributions. Second, the elemental or speciation distributions can be mapped by observing the spatial distributions of electron emissions across a surface. Thirdly, specific electron emissions to a predetermined depth and their depth distributions. This ranges from a few micrometers to less than 10 nm. Typically, analysis involves gathering energy spectra at all

4.5 X-Ray Photoelectron Spectroscopy (XPS)

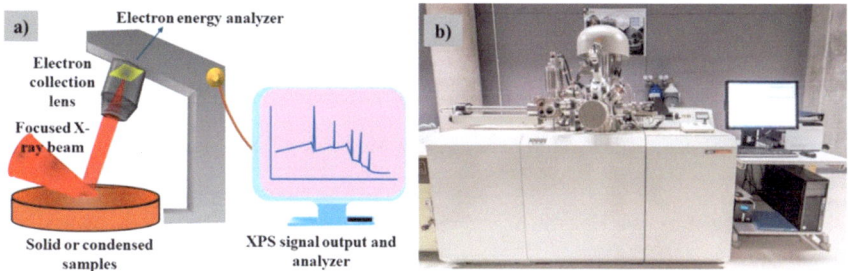

Fig. 4.9 a Fundamental components of monochromatic XPS function. **b** A photograph of XPS instrument (JEOL/JPS-9010MC model). Courtesy Thailand Scientific Equipment Center Network

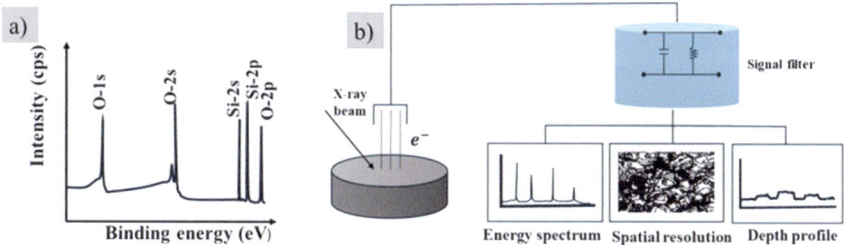

Fig. 4.10 a An illustration of gathered XPS spectra examined under Mg-Kα irradiation. **b** Fundamental XPS instrument components and the applicable data formats with energy spectrum, spatial resolution, and depth profile

available energies first, then focusing on photoelectron signals. This guarantees that every component is taken into consideration during quantification and that the data is gathered efficiently. The calculated binding energy (B.E.) is employed to create the energy spectrum, even though the detected kinetics energy (K.E.) is the quantity that is recorded in XPS. Since the B.E. is independent of the X-ray energy while the K.E. is dependent on it, the derived B.E. is utilized to create a spectrum. The Einstein equation establishes a relationship between the values of K.E., B.E., and the incident photon energy (E_i) [34]. Here, ϕ_{XPS} is the instrument's work function, not the work function of the specimen. This is included because, under the supposition that a conductive sample in physical contact with the instrument is being examined, it reflects the lowest energy required to remove one electron from the device [35].

$$\text{K.E.} = E_i - \phi_{XPS} - \text{B.E.} \tag{4.3}$$

References

1. G. Binnig, H. Rohrer, Scanning tunneling microscopy—from birth to adolescence. Rev. Mod. Phys. **59**(3), 615 (1987)
2. G. Binnig, H. Rohrer, In touch with atoms. Rev. Mod. Phys. **71**(2), S324 (1999)
3. K. Maeda, Y. Nakamura, Spreading effects in surface reactions induced by tunneling current injection from an STM tip. Surf. Sci. **528**(1–3), 110–114 (2003)
4. H.J. Zandvliet, A. van Houselt, Scanning tunneling spectroscopy. Annu. Rev. Anal. Chem. **2**(1), 37–55 (2009)
5. G. Missaoui et al., Chromium oxide thin films on Pt(111): an STM and DFT excursion through the phase diagram. J. Phys. Chem. C (2024)
6. J. Inukai et al., Direct STM elucidation of the effects of atomic-level structure on Pt(111) electrodes for dissolved CO oxidation. J. Am. Chem. Soc. **135**(4), 1476–1490 (2013)
7. G. Teobaldi et al., Role of applied bias and tip electronic structure in the scanning tunneling microscopy imaging of highly oriented pyrolytic graphite. Phys. Rev. B Condens. Matter Mater. Phys. **85**(8), 085433 (2012)
8. K. Hipps, Scanning tunneling spectroscopy (STS), in *Handbook of Applied Solid State Spectroscopy* (Springer, 2006), pp. 305–350
9. P.G. Collins et al., Scanning tunneling spectroscopy of C36. Phys. Rev. Lett. **82**(1), 165 (1999)
10. S. Wang et al., Automated tip conditioning for scanning tunneling spectroscopy. J. Phys. Chem. A **125**(6), 1384–1390 (2021)
11. A.M. Coe, G. Li, E.Y. Andrei, Quick connect scanning tunneling microscope head with nested piezoelectric coarse walkers. arXiv preprint arXiv:2403.01530 (2024)
12. Y. Zhou, J. Du, Atomic force microscopy (AFM) and its applications to bone-related research. Prog. Biophys. Mol. Biol. **176**, 52–66 (2022)
13. X. Shi et al., Atomic force microscopy-scanning electrochemical microscopy (AFM-SECM) for nanoscale topographical and electrochemical characterization: principles, applications and perspectives. Electrochim. Acta **332**, 135472 (2020)
14. M. Mahdavi et al., Modal actuation and sensing with an active AFM cantilever. IEEE Sens. J. **21**(7), 8950–8959 (2021)
15. P. Georgios, D. Philip, Real-time nanoparticle–cell interactions in physiological media by atomic force microscopy (2014)
16. P. Sudersan et al., Method to measure surface tension of microdroplets using standard AFM cantilever tips. Langmuir **39**(30), 10367–10374 (2023)
17. G. Stan, C.V. Ciobanu, S.W. King, Resolving the subsurface structure and elastic modulus of layered films via contact resonance atomic force microscopy. ACS Appl. Mater. Interfaces **14**(49), 55238–55248 (2022)
18. Q. Liu et al., Applications of AFM in membrane characterization and fouling analysis. ACS EST Eng. **4**(8), 1805–1838 (2024)
19. J.J. Schwartz, D.S. Jakob, A. Centrone, A guide to nanoscale IR spectroscopy: resonance enhanced transduction in contact and tapping mode AFM-IR. Chem. Soc. Rev. **51**(13), 5248–5267 (2022)
20. W. Xiang, Y. Tian, X. Liu, Dynamic analysis of tapping mode atomic force microscope (AFM) for critical dimension measurement. Precis. Eng. **64**, 269–279 (2020)
21. Y. Liu et al., Can AFM be used to measure absolute values of Young's modulus of nanocomposite materials down to the nanoscale? Nanoscale **12**(23), 12432–12443 (2020)
22. T. Lai, K. Shi, P. Huang, Adhesion force behaviors between two silica surfaces with varied water thin film due to substrate temperature studied by AFM. J. Adhes. (2020)
23. R. Pezone et al., Highly-sensitive wafer-scale transfer-free graphene MEMS condenser microphones. Microsyst. Nanoeng. **10**(1), 27 (2024)
24. C.-W. Yang et al., Imaging of soft matter with tapping-mode atomic force microscopy and non-contact-mode atomic force microscopy. Nanotechnology **18**(8), 084009 (2007)
25. K. Kobayashi, H. Yamada, K. Matsushige, Frequency noise in frequency modulation atomic force microscopy. Rev. Sci. Instrum. **80**(4) (2009)

26. G. Liu et al., Evolution of dip-pen nanolithography (DPN): from molecular patterning to materials discovery. Chem. Rev. **120**(13), 6009–6047 (2020)
27. Y. Li, B.W. Maynor, J. Liu, Electrochemical AFM "dip-pen" nanolithography. J. Am. Chem. Soc. **123**(9), 2105–2106 (2001)
28. S. Zaefferer, A critical review of orientation microscopy in SEM and TEM. Cryst. Res. Technol. **46**(6), 607–628 (2011)
29. X. Hao et al., From microscopy to nanoscopy via visible light. Light Sci. Appl. **2**(10), e108 (2013)
30. P. Song et al., Explicit analytical expressions for the electromagnetic field components of typical structured light beams. J. Quant. Spectrosc. Radiat. Transfer **241**, 106715 (2020)
31. D.N.G. Krishna, J. Philip, Review on surface-characterization applications of X-ray photoelectron spectroscopy (XPS): recent developments and challenges. Appl. Surf. Sci. Adv. **12**, 100332 (2022)
32. J.E. Whitten, Ultraviolet photoelectron spectroscopy: practical aspects and best practices. Appl. Surf. Sci. Adv. **13**, 100384 (2023)
33. S. Béchu et al., Photoemission spectroscopy characterization of halide perovskites. Adv. Energy Mater. **10**(26), 1904007 (2020)
34. J. Lefebvre et al., Experimental methods in chemical engineering: X-ray photoelectron spectroscopy-XPS. Can. J. Chem. Eng. **97**(10), 2588–2593 (2019)
35. M. Helander et al., Pitfalls in measuring work function using photoelectron spectroscopy. Appl. Surf. Sci. **256**(8), 2602–2605 (2010)

Chapter 5
Nanomaterial Synthesis and Characterization

Abstract One of the active transdisciplinary research areas in the intersections of solid-state physics, chemistry, biology, and engineering is the study of various nanomaterials. The abundance of synthetic materials found by industrialization and academia serves as evidence of this interest. One is that to keep lowering costs and speeding up the transmission and storage of information, new materials must be created on an ever-smaller scale to advance lab-on-a-chip sensors. Another is that, when compared to conventional materials, nanoparticles exhibit unique and frequently improved features, creating opportunities for new nanotechnological applications. Dendrimers used for self-assembly, which have numerous terminal functions and well-defined radial topologies, are the first material discussed in the opening section. They have a lot of potential for use in biomedical applications, such as sensors, diagnostics, drug administration, and therapies.

5.1 Bottom-Up and Top-Down Synthesis

The initial systems are assembled into ever more complex systems using the bottom-up approach shown in Fig. 5.1, thereby forming subsystems of the growing system. "Bottom-up processing" is a type of assembly process that builds a component by utilizing matter input from the surroundings. The essential components of the system are first fully specified using a bottom-up approach. These parts are then joined together to form larger subsystems, which are then connected periodically. In addition, top-down synthesis refers to an approach where bulk materials are disintegrated into nanostructures [1, 2].

Using the divergent growth method depicted in Fig. 5.2, dendrimers are produced by beginning with a multifunctional core and growing it outward through a series of processes. For example, the core molecule combines with a reagent containing reactive groups that serve various purposes to form a branch [3, 4]. Chemical vapor deposition (CVD) techniques are important in the bottom-up synthesis of carbon-based nanomaterials. In CVD, vapor-phase precursors react chemically to produce a thin coating on the substrate surface. If precursors have a long shelf life, good

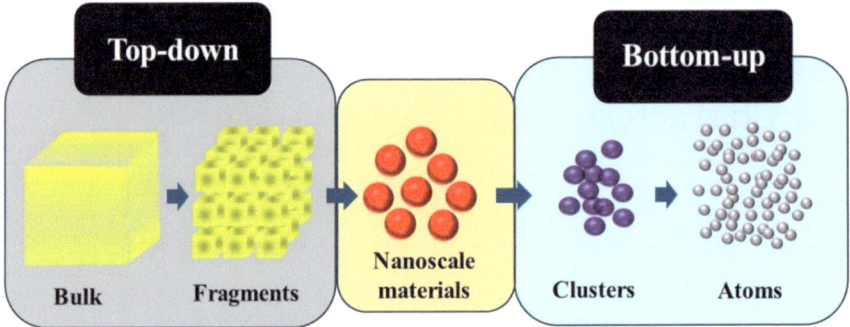

Fig. 5.1 Generic view of bottom-up and top-down synthesis

chemical purity, strong evaporation stability, low cost, and no harmful properties, they are considered suitable for CVD [5]. Its breakdown should also not leave any pollutants in its wake. For example, in the CVD method, a substrate is heated to high temperatures in an oven to produce carbon nanotubes. Next, a precursor gas containing carbon (such as hydrocarbons) is progressively added to the system. At high temperatures, the decomposition of the gas releases carbon atoms, which reunify to form carbon nanotubes on the substrate. However, the catalyst of choice has a major impact on the type and form of nanomaterial that is created. When the material is synthesized via CVD, Ni and Co catalysts create multilayer graphene, but a Cu catalyst only creates monolayer graphene [6]. For instance, with boron and nitride atoms in place of sublattices and a 1:1 stoichiometry between B and N, monolayer hexagonal boron nitride, or h-BN, is a geometric structural analogue of graphene. The unique ionicity of covalent B-N bonds gives it excellent thermal stability, electrical insulating properties, and acid–base resistance [7].

In essence, a top-down approach involves dissecting a system using reverse engineering to gain understanding of its compositional subsystems. It also goes by the names stepwise design, stepwise refinement, and decomposition in some contexts. A top-down approach is used to produce an overview of the system, identifying any first-level subsystems but without providing detailed information about them. Then, each subsystem is refined in even greater detail, possibly at several subsystem levels, until the entire specification is reduced to fundamental parts. "Black boxes" are often utilized to create a top-down model to facilitate manipulation. However, black boxes may not be detailed enough to evaluate the model in a realistic way or sufficiently depict basic mechanisms. A top-down approach starts with the big vision and then divides it into smaller, more doable pieces. To develop nanostructured materials, bulk components are separated using top-down techniques. Top-down methods include mechanical milling, electro-explosion, sputtering, etching, and laser ablation. Mechanical milling is a useful technique for producing compounds at the nanoscale from bulk materials. The method of mechanical milling shown in Fig. 5.3 is helpful for combining different phases and for producing nanocomposites. It is possible

5.1 Bottom-Up and Top-Down Synthesis

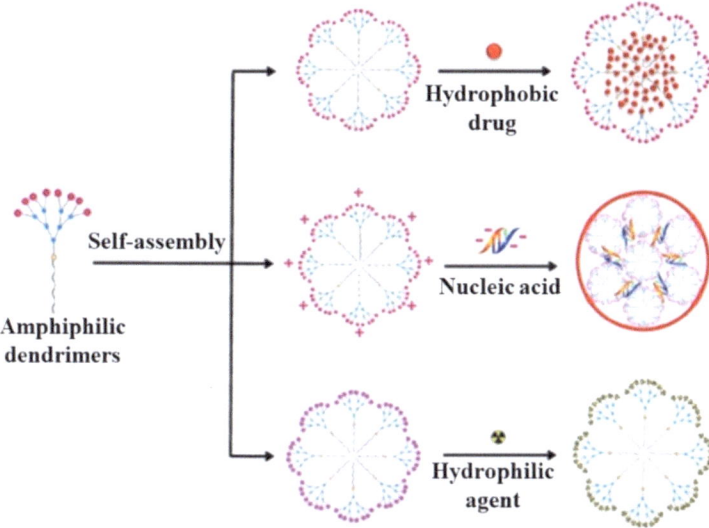

Fig. 5.2 Self-assembly of amphiphilic dendrimers into supramolecular dendrimers. Reproduced with a permission from Ref. [8]

to mechanically generate aluminum alloys enhanced by oxide and carbide, wear-resistant spray coatings, aluminum, nickel, magnesium, and copper nanoalloys, and a wide range of other nanocomposite materials [10, 11]. The phenomena known as sputtering in physics occurs when a solid substance is struck by energetic particles from a gas or plasma, causing tiny particles to be expelled from the material's surface. It is a naturally occurring occurrence in space and may cause unwanted wear on precise parts. It is used in science and industry, however, to perform precise etching, conduct analytical procedures, and deposit thin film layers in the production of optical coatings, semiconductor devices, and nanotechnology products. This is because it can be made to act on incredibly fine layers of material. It uses a physical method of vapor deposition. To produce nanomaterials by sputtering entails striking solid surfaces with high-energy particles, like gas or plasma. It is believed that sputtering is a practical method for producing thin films of nanomaterials. Strong gaseous ions strike the target surface during the sputtering deposition process, and depending on the incident gaseous-ion energy, they physically expel small atom clusters. There are several methods for sputtering, such as employing magnetrons, radio-frequency diodes, and DC diodes. Usually, sputtering happens in a chamber that has been emptied and filled with sputtering gas. A high voltage applied to the cathode target causes free electrons to collide with the gas, resulting in the formation of gas ions. The positively charged ions approach the cathode target and accelerate quickly in the electric field, ejecting atoms from the target's surface [12, 13].

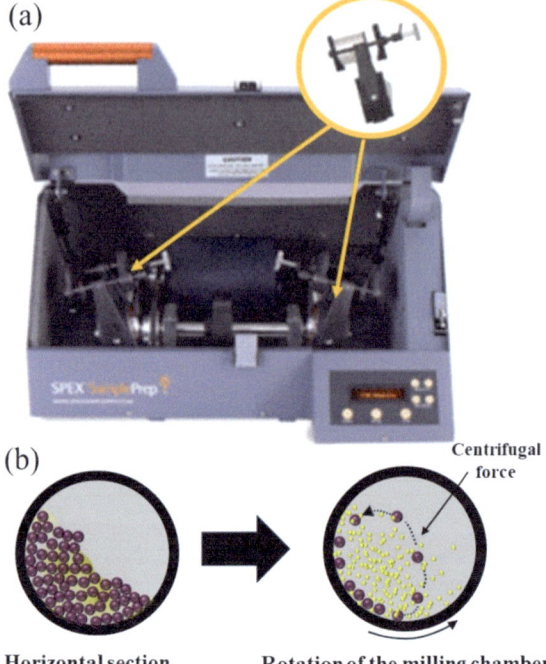

Fig. 5.3 a SPEX 8000 high-energy ball mill used for modeling analysis of ball-milling process for battery electrode. Reproduced with a permission from Ref. [9]. Copyright 2024 American Chemical Society. **b** Schematics showing grinding balls in the milling container moving in the lateral direction

5.2 Metal Fluoride Synthesis

The development of new material fabrication techniques is essential to the progress of science and technology. Because of fluorine's strong electronegativity, inorganic metal fluorides have outstanding chemical and physical properties that make them valuable for use in engineering goods. Scientists have discovered that nano-fluoride compounds, like the fluorinated nonporous cages in Fig. 5.4, have the intriguing physical characteristics. They abled to remove per- and polyfluoroalkyl substances (PFAS) that are accumulated in water resources. These substances POS serious environmental and health threats despite their nonbiodegradable behaviors [14]. Nevertheless, material scientists now face the difficulty of controlling the form, size, mono-dispersity, yield of the appropriate size, and other important features of nanocrystals with diverse chemistry. Figure 5.5 lists a variety of chemical compounds that contain fluorine, all of which have it in its initial oxidation state. Ionic or polar covalent bonds can be formed between atoms by fluorine. A more detailed description of the structural processes generated from isolated single crystals of real intermediates from reaction mixtures with different conversion levels may be found in Fig. 5.6. Metal alkoxides are typically aggregates rather than monomeric molecules. Whereas insoluble alkoxides form three-dimensional networks, soluble alkoxides exist as multinuclear clusters.

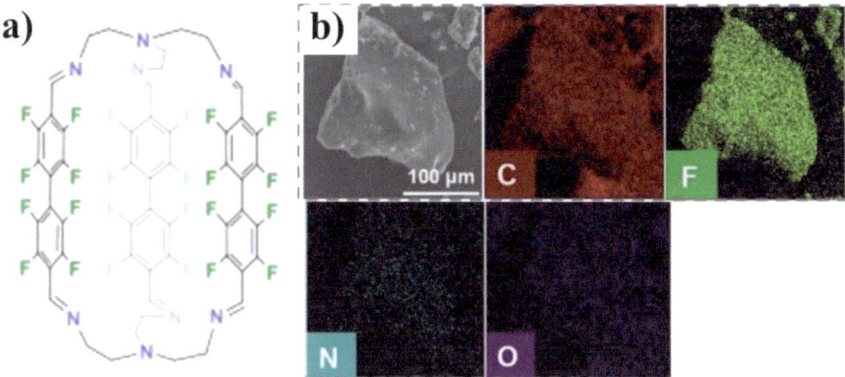

Fig. 5.4 **a** Schematic of a fluorinated nonporous adaptive crystalline cage. **b** SEM image and EDX mapping of the fluorinated cage. Reproduced with a permission from Ref. [14]. Copyright 2024 American Chemical Society

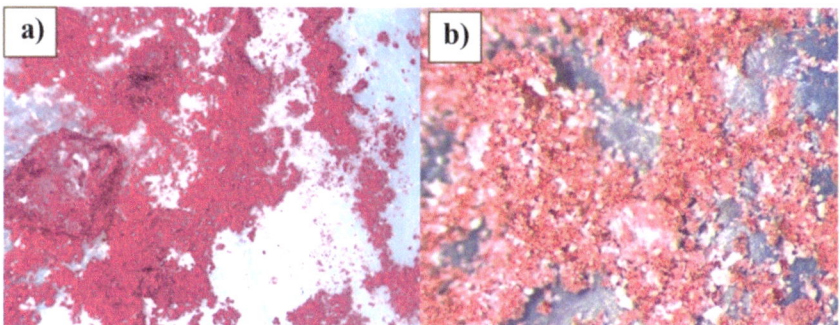

Fig. 5.5 The as-synthesized metal fluoride nanoparticles, such as **a** Mn_2F_5 and **b** Mn_3F_8 containing pale pink MnF_2. Reproduced with a permission from Ref. [15]. Copyright 2021 American Chemical Society

5.3 Ionic Liquid

Much attention has been focused on the use of ionic liquids (ILs), or organic salts that are liquid at room temperature, as a substitute for volatile organic solvents. Because they are nonflammable, nonvolatile, and recyclable, they are referred to as green solvents [17]. Aromatic cations that have been explored as ionic liquids, including the viologen-type, isoquinolinium, benzotriazolium, and pyridinium cations, as shown in Fig. 5.9. One carbon above the five-membered heterocyclic cations lies the second, albeit less well-known, class of heterocyclic room-temperature ionic liquids: the pyridinium RTILs. Although these salts have been known for some time, not as much attention has been paid to them as the imidazolium family has. This is most likely brought on by pyridine's toxicity and its reduced stability when nucleophiles

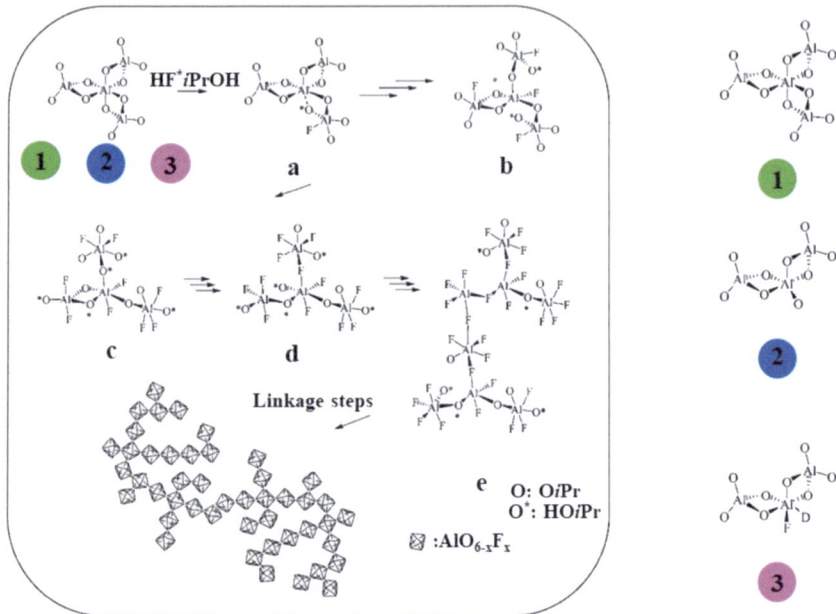

Fig. 5.6 Schematic representation of known structural elements occurring during the sol–gel fluorination of aluminum isopropoxide. Reproduced with a permission from Ref. [16]. Copyright 2009 American Chemical Society

are present. a variety of long-chain alkyl pyridinium hexafluorophosphate salts (C12–C18), some of which melt at temperatures lower than 100 °C. A relatively recent area of study is the viologen family of ionic liquids given in Fig. 5.7 [18] metathesis routes to ionic liquids. Although the melting solids of most viologens are very high, some do exhibit significantly lower melting temps, however they are still not quite room temperature. Benzotriazolium-based ILs are often efficient solvents for aromatic compounds. Using a halide or related salt of the desired cation, a metathesis reaction is used to prepare a huge range of ionic liquids. The general metathesis reaction can be divided into two groups based on the water solubility of the target ionic liquid: (1) metathesis via free acids or group 1 metals/ammonium salts, or (2) Ag salt metathesis. As electrolytes in high-energy electrochemical devices, ammonium-based ILs have been widely used because of their low melting temperatures, low viscosities, and superior electrochemical cathodic stabilities. These ILs work well as solvents and catalysts in a wide range of processes, but there are a few limitations on how they can be used. In the following sections, we will discuss the recognized advantages and disadvantages of each class of ILs [19].

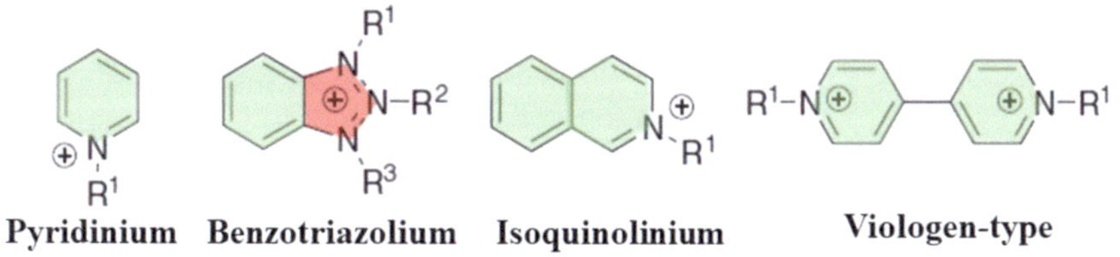

Fig. 5.7 Examples of different ionic liquid types

5.4 Transition Metal Dichalcogenides (TMDCs)

Modern synthetic methods and exfoliation techniques have made it possible to study and utilize the monolayers of layered transition metal dichalcogenides (TMDCs) depicted in Fig. 5.8. These TMDCs have long been of interest. Particularly significant are the naturally occurring minerals molybdenite and tungstite, or the compounds MoS_2 and WS_2, because they are both excellent at lowering friction and have inherent semi-conductivity. The anisotropic electrical and chemical properties of layered materials can be more pronounced due to their structure than in bulk semiconductors. The potential application of 2D materials for the catalysis of hydrogen evolution, hydrodesulfurization, and CO_2 reduction has also been studied because the edges of the sheets frequently operate as the active sites for a variety of processes. The relative number of the sites can be increased by modifying the morphology and, in the case of TMDCs, by doping [21]. TMDCs are often made using vapor deposition methods like physical vapor transport (PVT) or chemical vapor deposition (CVD). Since that the thickness of stacked TMDCs determines their optical and electrical properties, the process can be enhanced to regulate the size and number of layers in the materials. The method is adaptable and simple to change to include dopant species in the synthesis. These procedures usually involve the use of a powdered elemental form of chalcogen (S, Se, or Te) with either molybdenum oxide or tungsten oxide. Based on recent literature, the evolution of this method's extension through the addition of chalcogens and other metals is discussed below. Other wet chemical methods, including hydrothermal methods and the heating or hot injection of precursor solutions, are also employed in the production of TMDC. The limitations imposed by solvent boiling points make it more difficult to obtain greater temperatures in solution than in vapor phase processes, and there are additional factors to consider when evaluating the solubility compatibilities of reaction components. Solvent selection is one of the numerous parameters that offer more options to regulate the synthesis [22].

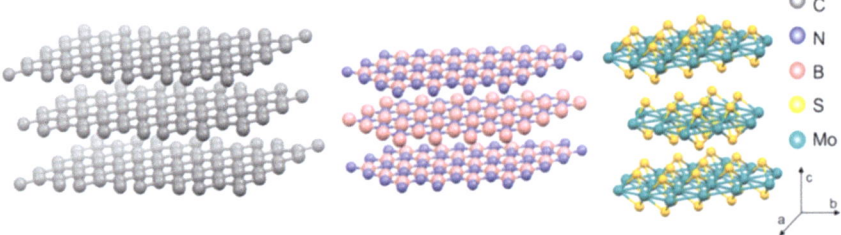

Fig. 5.8 Crystal structures of graphite and inorganic analogues of hexagonal boron nitride and molybdenum disulfide. Reproduced with Ref. [20]. Copyright 2016 CC-BY American Chemical Society

5.5 Green Synthesis of Nanoparticles

Many methods related to green chemistry have led to a large increase in the production of zinc and copper oxides as well as nanoparticles of metals such as gold, silver, and selenium in recent years. To enhance performance and safety in activities related to nanoparticle development, these nanoparticles are manufactured via green synthesis, as illustrated in Fig. 5.9. This technique is the most widely employed of the three methods for NP synthesis: physical, chemical, and biological. In contrast to bulk chemicals, the chemically manufactured material is either equally dangerous (as in the case of Ag) or less poisonous (as in the case of Au). Enzymes with

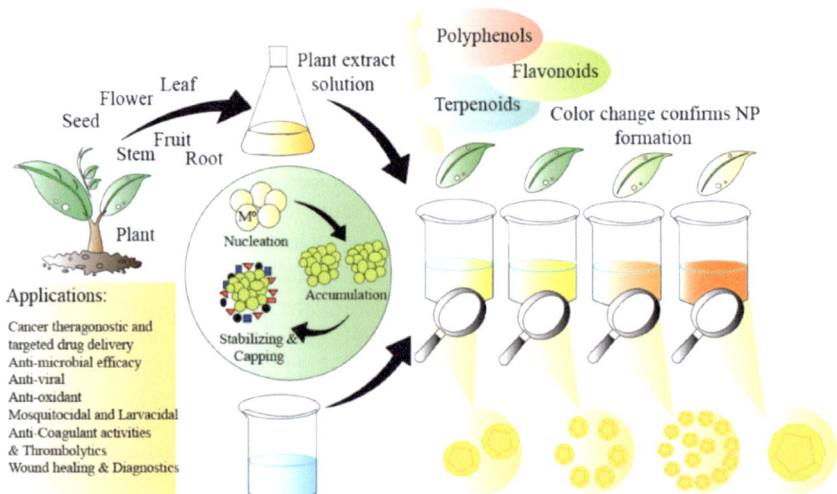

Fig. 5.9 Methodologies for the green synthesis of nanoparticles and their subsequent use in a wide range of applications. Reproduced with a permission from Ref. [23]. Copyright 2022 American Chemical Society

a well-defined structure and high availability are considered preferred for environmentally friendly synthesis methods. For example, Ag NPs were mixed using an enzyme-induced growth process on strong substrates during the NPs synthesis process. By exploiting electrostatic interactions, the enzymes were integrated into polymer multilayer-assembled membranes to facilitate the direct and environmentally friendly synthesis of bimetallic Fe/Pd particles within a membrane domain. Using ion exchange chromatography, a sulfatic reductase enzyme recovered from *Escherichia coli* was utilized to create a cell-free extract for the manufacture of Au NPs with antifungal activity against human pathogenic fungi. In a method of NP reduction using beetroot juice that was reported, the authors discovered that larger size Ag NPs were obtained by reducing the amount of beetroot juice. These NPs also demonstrated significantly higher catalytic activity and stability compared to those prepared with $NaBH_4$ for the conversion of 4-nitrophenol to 4-aminophenol [24]. Metal ions can be effectively chelated by a type of polyphenolic compounds called flavonoids to form nanoparticle. Their plethora of functional groups including hydroxyl or carbonyl group is believed to be responsible for their ability to generate NPs. The higher flavonoid and phenol levels of the aqueous extract of *Rumex dentatus* facilitated the bio-reduction of Ag^+ to Ag^0. The keto-enol tautomeric transition of the flavonoid has been proposed to be the cause of the release of reactive hydrogen atoms, which leads to metal ion reduction. For example, extracts of *Ocimum basilicum* contain the flavonoids luteolin and rosmarinic acid, which have been shown to play a major role in the production of NPs in the keto-enol form [25].

References

1. N. Abid et al., Synthesis of nanomaterials using various top-down and bottom-up approaches, influencing factors, advantages, and disadvantages: a review. Adv. Coll. Interface Sci. **300**, 102597 (2022)
2. Y. Wang, Y. Xia, Bottom-up and top-down approaches to the synthesis of monodispersed spherical colloids of low melting-point metals. Nano Lett. **4**(10), 2047–2050 (2004)
3. D.R. Sikwal, R.S. Kalhapure, T. Govender, An emerging class of amphiphilic dendrimers for pharmaceutical and biomedical applications: Janus amphiphilic dendrimers. Eur. J. Pharm. Sci. **97**, 113–134 (2017)
4. A.P. Schenning et al., Amphiphilic dendrimers as building blocks in supramolecular assemblies. J. Am. Chem. Soc. **120**(32), 8199–8208 (1998)
5. H. Frey, Chemical vapor deposition (CVD), in *Handbook of Thin-Film Technology* (2015), pp. 225–252
6. Y. Nakajima et al., Metal catalysts for layer-exchange growth of multilayer graphene. ACS Appl. Mater. Interfaces **10**(48), 41664–41669 (2018)
7. K.K. Kim et al., Synthesis of monolayer hexagonal boron nitride on Cu foil using chemical vapor deposition. Nano Lett. **12**(1), 161–166 (2012)
8. Z. Lyu et al., Self-assembling supramolecular dendrimers for biomedical applications: lessons learned from poly(amidoamine) dendrimers. Acc. Chem. Res. **53**(12), 2936–2949 (2020)
9. C. Welch, K.T. Cho, V. Srinivasan, Modeling analysis of ball-milling process for battery-electrode synthesis. Chem. Mater. **36**(14), 6748–6764 (2024)
10. H. Ghayour, M. Abdellahi, M. Bahmanpour, Optimization of the high energy ball-milling: modeling and parametric study. Powder Technol. **291**, 7–13 (2016)

11. V. Šepelák, S. Bégin-Colin, G. Le Caer, Transformations in oxides induced by high-energy ball-milling. Dalton Trans. **41**(39), 11927–11948 (2012)
12. J. Liang et al., Magnetron sputtering enabled sustainable synthesis of nanomaterials for energy electrocatalysis. Green Chem. **23**(8), 2834–2867 (2021)
13. J.T. Gudmundsson, Physics and technology of magnetron sputtering discharges. Plasma Sources Sci. Technol. **29**(11), 113001 (2020)
14. Y. He et al., Fluorinated nonporous adaptive cages for the efficient removal of perfluorooctanoic acid from aqueous source phases. J. Am. Chem. Soc. **146**(9), 6225–6230 (2024)
15. J. Bandemehr et al., Syntheses and characterization of the mixed-valent manganese (II/III) fluorides Mn_2F_5 and Mn_3F_8. Inorg. Chem. **60**(17), 12651–12663 (2021)
16. R. König, G. Scholz, E. Kemnitz, Local structural changes in aluminum isopropoxide fluoride xerogels and solids as a consequence of the progressive fluorination degree. J. Phys. Chem. C **113**(16), 6426–6438 (2009)
17. M. Grabda et al., Theoretical selection of most effective ionic liquids for liquid–liquid extraction of NdF_3. Comput. Theor. Chem. **1061**, 72–79 (2015)
18. M. Honda et al., Synthesis of novel fluorous pyridinium ionic liquids. Heteroat. Chem. **29**(5–6), e21464 (2018)
19. K. Ghandi, A review of ionic liquids, their limits and applications. Green Sustain. Chem. **2014** (2014)
20. A.A. Tedstone, D.J. Lewis, P. O'Brien, Synthesis, properties, and applications of transition metal-doped layered transition metal dichalcogenides. Chem. Mater. **28**(7), 1965–1974 (2016)
21. W. Choi et al., Recent development of two-dimensional transition metal dichalcogenides and their applications. Mater. Today **20**(3), 116–130 (2017)
22. Y. Zhang et al., Recent progress in CVD growth of 2D transition metal dichalcogenides and related heterostructures. Adv. Mater. **31**(41), 1901694 (2019)
23. Y. Yao et al., Microconfinement from dendronized chitosan oligosaccharides for mild synthesis of silver nanoparticles. ACS Appl. Nano Mater. **5**(3), 4350–4359 (2022)
24. A. Gour, N.K. Jain, Advances in green synthesis of nanoparticles. Artif. Cells Nanomed. Biotechnol. **47**(1), 844–851 (2019)
25. B. Teofilović et al., Analysis of functional ingredients and composition of *Ocimum basilicum*. S. Afr. J. Bot. **141**, 227–234 (2021)

Chapter 6
DNA Nanotechnology

Abstract The unique topic of deoxyribonucleic acid (DNA) nanotechnology can be guided and advanced by the multidisciplinary subject that involves the advancement of physics, chemistry, biology, mathematics, engineering, and materials science. For several decades that have passed since Nadrian Seeman's initial suggestion of DNA origami, remarkable progress has been at a center of interest for development to healthcare applications. In living systems, genetic data is stored and transmitted via the fundamental molecule, recognized as DNA. DNA nanotechnology extracts this molecular information from its biological environment and utilizes its data to put together structural motifs and then fabricate or link them together. This area of study and research has significantly evolved nanotechnology and nanoscience and improved our understanding of molecular self-assembly over decades.

6.1 Role of DNA

Molecular self-assembly plays a crucial role in the intricate architecture and functionality of biological systems. At the micro, nano, and atomic levels, nature has created multipart ways for information to self-assemble and move certain biomaterials into ordered cellular frameworks. To be more precise, we can think of the cell as a biological factory where various molecular equipment works in tandem. Synthetic biology aims to create multifunctional biological mimics that enable artificial signaling and communication in a fully regulated manner. DNA chains are better suited than proteins or other biomolecules for a variety of purposes, including the creation of functional materials and biological mimics through molecular self-assembly. First off, DNA is a clearly defined double helix with a right-handed structure composed of two complementary single strands with a diameter of around 2 nm and 10.5 bases that are 3.4 nm long for each helical turn. The fundamental building block of all nucleic acid nanostructures is the double helix. Therefore, knowledge of the double helix's structure and physicochemical properties is essential when creating ordered assemblies of multiple DNA units and using them for additional functionalization. The polymeric DNA can be formed by periodically repeating four distinct nucleotides. Each nucleotide is composed of a phosphate group, the five-carbon sugar deoxyribose, and a nitrogenous base, also referred to as a nucleobase or just a base. The four nucleobases are adenine (A), guanine (G), thymine (T), and cytosine (C) [1, 2]. The sugar-phosphate moieties provide the structure's backbone, while the four bases are responsible for giving the nucleic acid helix its helical shape. But hydrogen bonds need to form between each strand's bases for the double-stranded (ds) form to form a stable double helix. Base pairing is the process where two nucleobase pair together (T and G pair, A pairs with C), because each nucleobase only couples with one other nucleobase, referred to as the complement. Any DNA design is fundamentally based on a Watson–Crick base-pairing rule that allows the unique identification of a complementary DNA sequence given a certain DNA sequence. The polynucleotide chain often contains genetic information necessary for an organism's or virus's growth, development, and reproduction [3, 4].

6.2 DNA Motifs

The process of self-assembling on DNA nanostructures or motifs starts with directing multiple double helices to create what are referred to as branched DNA motifs, or DNA tiles. The Holliday junction, a crucial nucleic acid connection in numerous DNA recombination and double strand break repair mechanisms, is the most fundamental motif. When two homologous double-strand DNA molecules exchange strands at a central branch point, the Holliday junction is created. As such, the branch point's motion is mostly dictated by the symmetry of the sequences that comprise a hypothetical Holliday junction. Depending on the sequence of nucleobases nearest to the junction and the concentration of buffer salts, the DNA arms may take on one of several conformations. Although this is obviously necessary for the biological function to perform as intended, it necessitates a high degree of fabrication precision because a particular level of structural rigidity is required for each of the individual building blocks. The original goal was to create a stationary Holliday junction to prevent topological isomerization and prevent the motif from splitting into two separate double-helical segments. In a different important class of nanostructures, the DNA motif is simply recognized by the entering proteins without being broken down or processed. This class of nanostructures as a bottom-up nanofabrication approach is useful for molecular tags such as those that are used to target transcription factors or position DNA-binding protein adaptors at specified sites. For biosensing applications, a wide family of functional DNA nanostructures is made of DNA aptamers [6]. As shown in Fig. 6.1a, the design of protype DNA origami triangles with none of hairpins (0HP) and 258 hairpins (258HP) used as scaffolds. Obtained TEM images of purified 258HP and 0HP origami triangles are demonstrated in Fig. 6.1b. The origami fabrication was relied on extending triplex directed functionalization and cross-linking from DNA tiles to DNA origami. The triplex-forming oligonucleotides integrating with hairpins enhanced the structural integrity of DNA origami and make the nanostructure less susceptible to chaotropic agents.

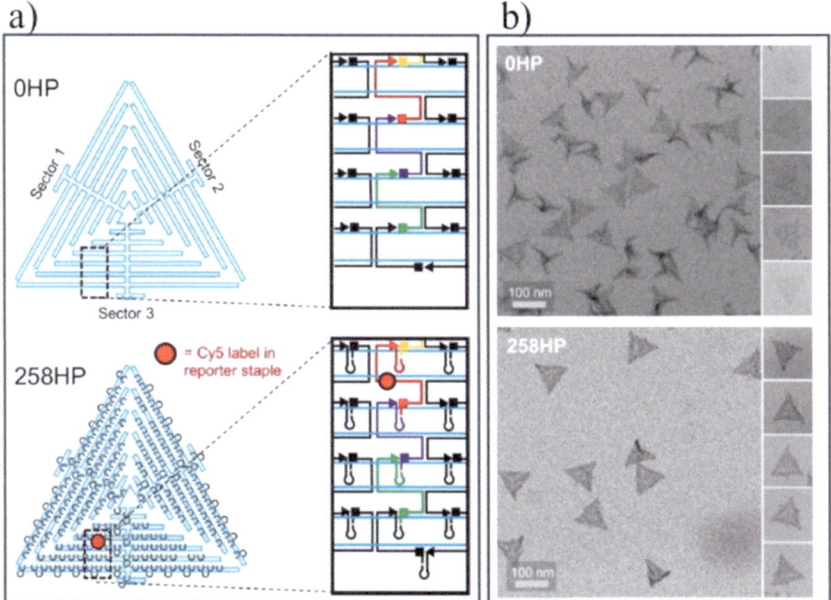

Fig. 6.1 a Schematic of protype DNA origami triangles with and without the molecular hairpins. **b** TEM images of purified 258HP and 0HP origami triangles. Reproduced with a permission from Ref. [5]. Copyright 2024 CC-BY 4.0 American Chemical Society

6.3 Wireframe DNA Origami

The original DNA-origami design features the scaffold strand and associated double helices arranged into raster-fill patterns that are parallel, as shown in Fig. 6.2. As a result, structures are densely packed and comprise domains of adjacent antiparallel helices connected by double crossover DNA links or DX motifs. Wireframe structures that enable non-parallel orientation of DNA helices can be created by connecting multi-arm junction units, which have two to twelve arms each [7]. This yields numerous layers of frameworks, three-dimensional polyhedrons, curved solids, and two-dimensional network-like surfaces with voids of different sizes and shapes. DNA segments were able to successfully fold the scaffold into planar gridiron shapes by employing a fundamental repeating unit called a four-arm junction. Experimental investigations revealed that rhomboid structures rather than square ones were produced because of the motif's flexibility and relaxed conformation into a right-handed twist with a 60° torsion angle. The process was expanded to three dimensions by stacking or weaving many layers of two-dimensional gridiron lattices at specific connecting points, or by adjusting the curvature of the individual motifs to create distorted patterns. Utilizing a special layered-crossover motif called the LX motif, multilayers of DNA frameworks with exact geometry, long-term mechanical stability, and modifiable cavity size have recently been produced. The LX motif is

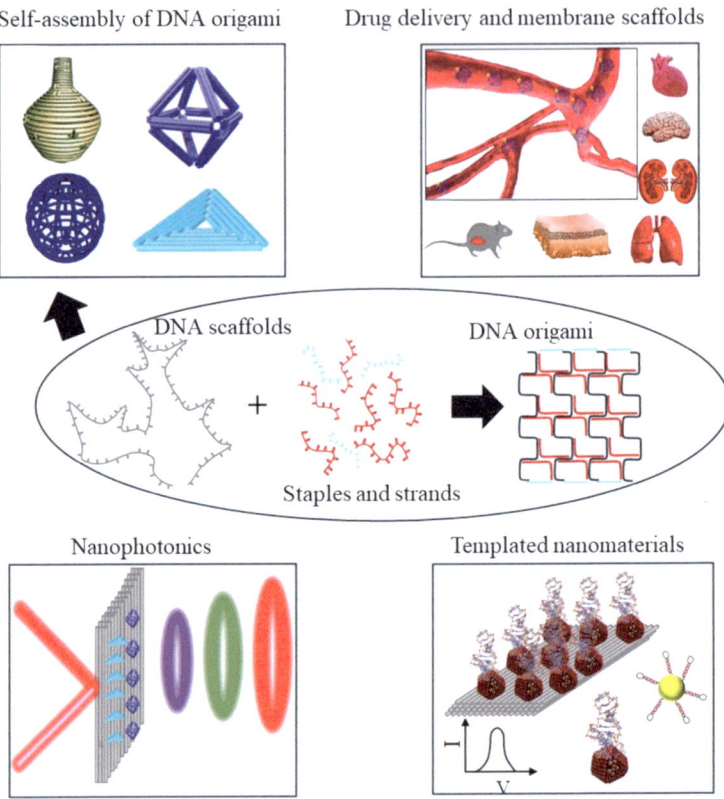

Fig. 6.2 Wireframe-based DNA origami designs using different engineered nanomaterials for diverse applications

tailored from the DX motif, but instead of connecting DNA helices in the same plane as those in the DX motif, its crossings correspond to two parallel layers. As a result, DNA arms can be positioned non-parallel across layer stacks, increasing the structural DNA nanotechnology toolbox. By constructing randomly generated links between multiple arm connectors in two and three dimensions, the same group illustrated the idea of adaptability. The four-step procedure is the foundation of the DNA origami design. Every bridging between a vertex in the wireframe structure is changed to a double line in the first step. The second step involves looping and bridging lines so that the scaffold strand passes through each vertex just once and allows a circular path to be built. A third step creates DX tiles divided by the total number of helical turns by relating the scaffold lines to the corresponding staple strands. This releases the tension and allows the scaffold to travel along its predetermined path. Changing the angles of vertices is the last step. This is achieved by leaving a specific number of nucleobases unpaired in the scaffold on the opposite side of the Tn loop and introducing a nucleic acid loop of the required length into the staple strands around the vertex. To enable the arms to bend and arrange themselves in the most advantageous way possible—thereby producing the ideal combination of angles—each vertex is given a certain amount of structural flexibility. This approach has been used to generate complicated, arbitrary designs in the shape of three flowers and a bird, as well as simple platonic tiling and curved patterns. Schlegel diagrams have also been employed in three-dimensional Archimedean symmetry systems to build the scaffold's looping path [4, 8–10].

6.4 Engineering Functional DNA–Protein Conjugates

The development of DNA–protein conjugation chemistries that are incredibly efficient, robust, and dependable as theragnostic agents (see Fig. 6.3) is essential for its applications [11–13]. The design principle of an idea synthesis approach should have the following qualities: The benefits of this approach are as follows: (1) maintenance of the natural activities of the protein; (2) excellent yield, high efficiency, and cheap cost; (3) simple synthesis without time-consuming DNA alterations; and (4) stable and robust chemistry under physiological settings. One of the most used conjugation methods, the biotin–avidin interaction, was used to immobilize DNAs onto protein substrates. The biotin–avidin connection, one of the strongest non-covalent bindings, shows remarkable stability even in the face of chemical and heat denaturation. The binding affinity between avidin and biotin, which is around 10^{-12} M, can be increased by using streptavidin (STV), which has a K_a of about 10^{-14} M, and neutravidin, which has a K_a of about 10^{-16} M. During synthesis, the target protein's hydrophobic avidin epitope is fused to it, and biotin is employed to modify the 3′ or 5′ ends of the target DNA strands. Another often used tactic is antibody (Ab)-antigen interaction. Haptens such as digoxigenin, dinitrophenol, and FLAG tag are commonly used to modify DNA terminals so that they can bind to more related antibodies. One of the main advantages of using the Ab-antigen interaction to create DNA–protein conjugates is the hapten's (small molecule binding motif) easy incorporation into the DNA solid-state synthesis process.

Aptamer, a monoclonal Ab congener, is a kind of single-stranded oligonucleotide that binds to target proteins with high affinity and specificity by folding into three-dimensional binding pockets to be recognized [14]. Therefore, DNA aptamer has also been used to prepare DNA–protein conjugate through a specific interaction. A DNA sequence with two segments is produced when employing an aptamer to attach to a conjugate protein; one segment carries out DNA hybridization while the other functions as an aptamer and binds to the target protein. One well-liked aptamer sequence that has been used in many diagnostic assays is anti-thrombin aptamer. The supply of aptamers for desired targets is decreased by the time-consuming in vitro selection process involved in identifying new aptamers for target proteins, even though using aptamers has many advantages, such as enhanced chemical stability and ease of modification. Furthermore, the aptamer's average binding affinity to the protein is only around 10^{-7}–10^{-8} M, which is insufficiently stable for use in in vivo settings. A

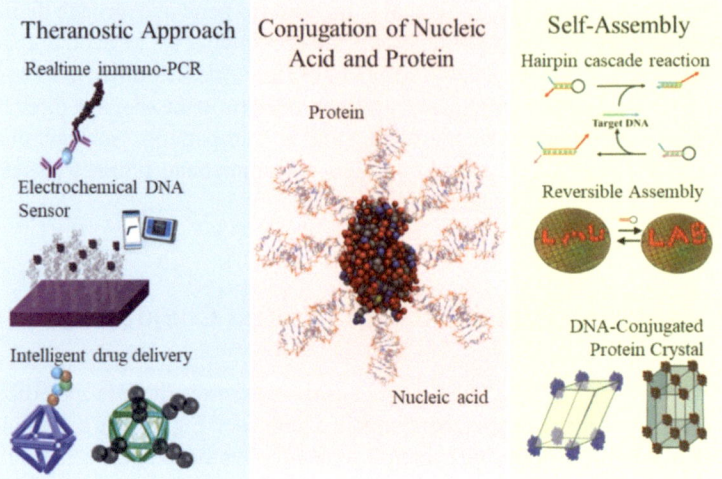

Fig. 6.3 Schematic use of nucleic acid-protein conjugates as theragnostic agents and component parts in a variety of applications

traditional biosensor's general architecture consists of signal transduction motifs and target identification. The target recognition motif, such as an antibody or an aptamer, recognizes biomolecules of interest with high selectivity and binding affinity, whereas the signal transduction motif translates the binding process into detectable signals and subsequently amplifies those signals to obtain high sensitivity. Proteins and DNA have different biological functions, which allow them to be utilized in different contexts for signal transduction, amplification, and identification. Consequently, two groups of DNA–protein conjugate-based biosensors for diagnostic applications may be distinguished based on the diverse roles that DNA and protein play. This section looks at the two types of biosensors: those that use proteins as recognition motifs and DNA as signaling motifs. Since the methods by which functional DNAs and antibodies recognize biotargets have already been well researched, here we will focus on how DNAs and enzymes transduce and amplify signals in a typical biosensor design to offer a low enough limit of detection.

6.5 Bioassays

Terms like "PCR" (polymerase chain reaction) and "antigen rapid diagnostic test" (RDT) are no longer only used in scientific contexts but are now frequently used in everyday life. This is because common clinical diagnostic procedures for infectious diseases have become more accessible to the public since the COVID-19 pandemic. Lateral flow biosensors use RDTs that are paper based. Because these tests are low-cost, portable, easy to administer, and require no equipment or batteries, they are highly suitable for POCT, low-resource settings, and emergency use [16]. In brief, capillarity action serves as the physical foundation for the lateral flow (LF) testing, as shown in Fig. 6.4. When the extracted material is placed in the sample pad, the conjugate pad contains the nanoparticles (NPs) conjugated with the bioreceptor, ready to react with the target analyte (if present). If the specimen includes the target, the flow will appear at the detecting pad (membrane) and the test line (TL) will appear, indicating a positive result due to the presence of a specific bioreceptor that identifies the target. Therefore, if there is no target in the sample, the TL will not be shown. As the sample continues to flow across the detecting pad, a control line (CL) with a secondary bioreceptor ought to be visible for all trustworthy tests. At the end of the strip, an absorbent pad is positioned to catch any excess fluid. Aptamers are single-stranded DNA or RNA molecules that bind to a target with great affinity and selectivity. They fold into complex 2D or 3D structures. They perform significantly

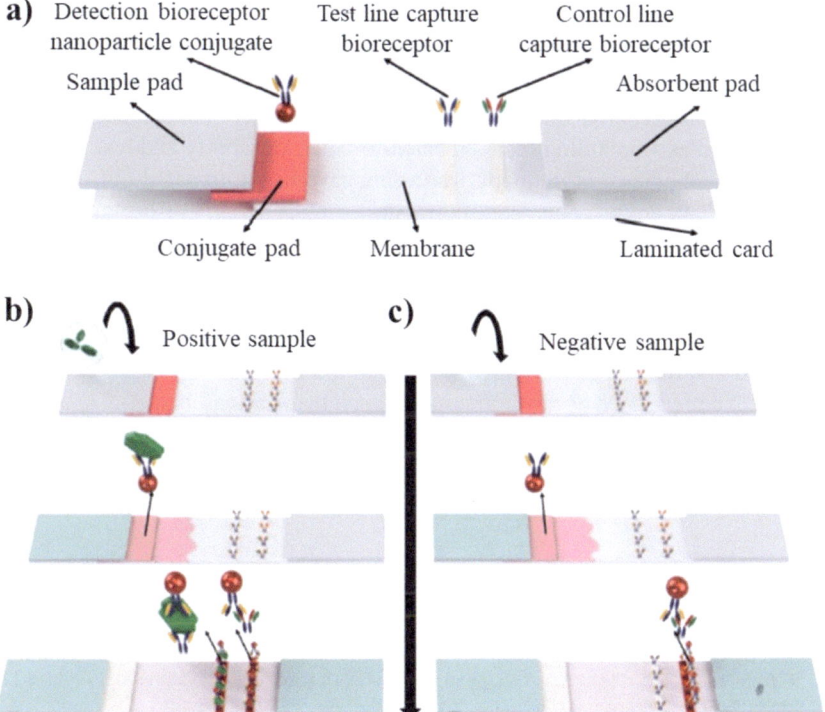

Fig. 6.4 An LF test's components and operation are demonstrated. **a** An LF strip consists of four main parts: the sample pad, conjugate pad, membrane (which prints the bioreceptors), and absorbent pad. All these parts are placed on a card that has been laminated. A conventional LF immunoassay's operation (**b, c**). **b** The aggregation of nanoparticles on TL and CL results in the creation of two red lines when the target is present in the sample. **c** Only the CL is displayed when the target is not present. Reproduced with a permission from Ref. [15]. Copyright 2022 American Chemical Society

better than Ab in several aspects, such as large-scale synthesis capability, in vitro production, functionalization simplicity, and high stability. The systematic evolution of ligands by exponential enrichment (SELEX) test-tube procedure, which has been adjusted according to the target of interest, is the source of aptamers. This approach bears similarities to artificial selection, controlled evolution, and natural selection. SELEX or its derivatives have generated aptamers for hormones, drugs, heavy metals, proteins, toxins, enzymes, cancer cells, and bacterial, viral, and fungal infections. Aptamers are used as bioreceptors in aptamer-based LF testing; however, they are not able to address the LF's lack of sensitivity, and their ability to bind is strongly influenced by the buffer's ionic strength and the presence of specific cations [17, 18].

The similar idea underlies the multiplexed LF assays, which allow for the simultaneous detection of multiple targets. Printing each TL with its corresponding bioreceptor in the detecting pad is the simplest way to create multiplexed LFs. However, the maximum number of targets that may be multiplexed with this method is limited by the detection pad's length and/or the range of labels that are available. To address these problems, new methods and technologies have recently been developed (e.g., fluidic devices, microarrays, or the use of several NPs in the same test). The most utilized bioreceptors in LF immunoassays (LFIAs) are monoclonal and polyclonal antibodies. But recently, due to several advantages over monoclonal and polyclonal antibodies—including reduced batch-to-batch variability and less unspecific binding—nanobodies and antibody fragments have begun to stand out. The usual targets of LFIAs are proteins (e.g., hormones, antigen viruses or bacteria, toxins), while alternative targets such as drugs of abuse, cofactor chemicals, or even heavy metals can also be recognized. Ab has several drawbacks despite its extensive application in LF, including significant batch-to-batch variability, a lengthy development period, the need for ethical approval, and the requirement for refrigerated storage. Furthermore, it may be challenging to find high-affinity antibodies for some nonimmunogenic targets, such as tiny compounds (such medications, bacterial spores, and insecticides). Aptamers, sometimes known as "chemical Ab," have been studied as LF bioreceptors to overcome these limitations imposed by the Ab [19–21].

References

1. N.C. Seeman, H.F. Sleiman, DNA nanotechnology. Nat. Rev. Mater. **3**(1), 1–23 (2017)
2. R. Rohs et al., The role of DNA shape in protein–DNA recognition. Nature **461**(7268), 1248–1253 (2009)
3. I.I. Cisse, H. Kim, T. Ha, A rule of seven in Watson–Crick base-pairing of mismatched sequences. Nat. Struct. Mol. Biol. **19**(6), 623–627 (2012)
4. P. Zhan et al., Recent advances in DNA origami-engineered nanomaterials and applications. Chem. Rev. **123**(7), 3976–4050 (2023)
5. S. Kalra et al., Functionalizing DNA origami by triplex-directed site-specific photo-cross-linking. J. Am. Chem. Soc. **146**(19), 13617–13628 (2024)
6. O.I. Wilner, I. Willner, Functionalized DNA nanostructures. Chem. Rev. **112**(4), 2528–2556 (2012)
7. S. Dey et al., DNA origami. Nat. Rev. Methods Primers **1**(1), 13 (2021)

References

8. A.-T. Phan, M. Guéron, J.-L. Leroy, Investigation of unusual DNA motifs. Methods Enzymol. **338**, 341–371 (2002)
9. F. Touzain et al., DNA motifs that sculpt the bacterial chromosome. Nat. Rev. Microbiol. **9**(1), 15–26 (2011)
10. Y. Dong, Z. Yang, D. Liu, DNA nanotechnology based on i-motif structures. Acc. Chem. Res. **47**(6), 1853–1860 (2014)
11. B.B. Mendes et al., Nanodelivery of nucleic acids. Nat. Rev. Methods Primers **2**(1), 24 (2022)
12. J.A. Kulkarni et al., The current landscape of nucleic acid therapeutics. Nat. Nanotechnol. **16**(6), 630–643 (2021)
13. J. Baranwal et al., Electrochemical sensors and their applications: a review. Chemosensors **10**(9), 363 (2022)
14. D.S. Seferos et al., Polyvalent DNA nanoparticle conjugates stabilize nucleic acids. Nano Lett. **9**(1), 308–311 (2009)
15. A. Rubio-Monterde, D. Quesada-González, A. Merkoçi, Toward integrated molecular lateral flow diagnostic tests using advanced micro- and nanotechnology. Anal. Chem. **95**(1), 468–489 (2022)
16. Y. Zhou et al., Point-of-care COVID-19 diagnostics powered by lateral flow assay. TrAC Trends Anal. Chem. **145**, 116452 (2021)
17. H. Dong et al., Combination of capture-SELEX and post-SELEX for procymidone-specific aptamer selection and broad-specificity aptamer discovery, and development of aptamer-based lateral flow assay. Anal. Chim. Acta **1318**, 342922 (2024)
18. E.B. Bahadır, M.K. Sezgintürk, Lateral flow assays: principles, designs and labels. TrAC Trends Anal. Chem. **82**, 286–306 (2016)
19. F. Di Nardo et al., Ten years of lateral flow immunoassay technique applications: trends, challenges and future perspectives. Sensors **21**(15), 5185 (2021)
20. L. Zhan et al., The role of nanoparticle design in determining analytical performance of lateral flow immunoassays. Nano Lett. **17**(12), 7207–7212 (2017)
21. A. Van Amerongen et al., Lateral flow immunoassays, in *Handbook of Immunoassay Technologies* (Elsevier, 2018), pp. 157–182

Chapter 7
Development of Wearable Sensors for Sensing Applications

Abstract Today's healthcare systems are primarily overactive. Once symptoms appear, patients seek medical attention from doctors, who then treat and observe them in a passive capacity. This method, which prioritizes diagnosis and treatment over preventive healthcare, largely fails to halt the onset of illnesses. It further prevents people from actively participating in their own health monitoring. By enabling people to understand the workings of their own physiology, the rapidly expanding field of portable biosensors, wearable sensors, or wearables seeks to overcome the shortcomings of overactive and centralized healthcare. The long-term goal is to create sensors that may be incorporated into wearable forms, such as wristbands, patches, tattoos, or clothes, to continuously monitor a variety of bodily signs. These sensors will allow people to monitor themselves without costly equipment or qualified professionals by transmitting physiological information as the body changes between healthy and unhealthy states. Commercial wearable devices now incorporate digitalized telemedicine technologies with the Internet of Things (IoT) for noninvasive remote health monitoring. These tools are becoming more and more integrated into our daily lives. A new era of decentralized healthcare has been ushered in with the newly established tools. For any industry involved in healthcare, the development of a wearable monitoring device for telemedicine has generated a great deal of interest in the multimodal big data capture of real-time physiological and biochemical data using non-invasive techniques. The goal of telemedicine wearable development has been to minimize the cost of invasive and high-tech therapies while enabling early identification of a variety of ailments. The advancement of biosensors toward the production of wearable sensors opens the chapter.

7.1 Components of Biosensors

Analyzing biochemical and physicochemical processes is crucial for a variety of applications in engineering, biotechnology, medicine, and nanotechnology. However, the challenge of directly connecting an electronic device to a biological environment necessitates the translation of biological understanding into a readily understandable digital signal. Since electrochemical biosensors directly transform a biological event into a readable signal, they provide an intriguing window into the analysis of biological samples [1]. Many sensing methods and associated technologies have emerged in the last few decades. Not only do promising results emerge from interesting new areas such as biosensing with aptamers, peptides, or even magnetic nanoparticles, but also from widely used conventional techniques like impedance spectroscopy, chronoamperometry, cyclic voltammetry, and chronopotentiometry. Additionally, there are other methods such as electrochemical sensing platforms, quartz crystal microbalance, ellipsometry, surface plasmon resonance, optical waveguide light mode spectroscopy, and scanning probe microscopy [2–4]. These techniques have been demonstrated to function well in conjunction with electrochemical detection through the use of the common biosensing modules depicted in Fig. 7.1. Often, surface modification and electrical interface designs that create a nanoscale link between the sensing device and the target sample control the signal transduction and overall performance of electrochemical sensors. The most popular methods of surface functionalization, different electrochemical transduction systems, and the selection of recognition receptor molecules all influence the sensor's final sensitivity. More alternatives for signal amplification are provided by novel nanotechnology-advanced techniques including encapsulating enzymes in vesicles, polymersomes, or polyelectrolyte capsules or using designed ion-channels in lipid bilayers. The field of biosensors research has shown exponential growth in the last twenty years. A biosensor is often an analytical instrument that converts a biological response into a signal that can be measured and used for practical purposes [5–7].

For developing an efficient biosensor for the non-specialist market, several conditions must be met. For target detection, the biocatalyst or artificial nanomaterials must be stable under normal storage settings, exhibit very little fluctuation between experiments, and have specificity and selectivity. Temperature, pH, and stirring must have the least possible impact on the reaction. Consequently, samples require little to no pre-treatment to analyze them. If the process requires cofactors or coenzymes, it is preferable to immobilize them alongside the enzyme. Throughout the concentration range of interest, the signal detection should be exact, precise, reproducible, and linear. Additionally, it must be free of transducer-related signal–noise interference from electrical and other sources [10]. If the portable or point-of-care (POC) biosensor is used for non-invasive monitoring in clinical settings, the probe or electrical electrode must be small, flexible, biocompatible, and devoid of negative or allergic consequences. Moreover, the biosensing materials should not be vulnerable to heat stimuli or proteolysis. Quick detection of target analytes from human samples are best achieved through real-time analysis from the biosensors. The whole biosensors ought to be reasonably priced, durable, lightweight, portable, and suitable for semi-skilled users. Biosensors should be very selective and repeatable as they can precisely control the interaction of chemicals by immobilizing biological recognition components on the sensor substrate that have a specific binding affinity to the desired molecule. Biological receptors, whole cells, nucleic acids, amino acids, enzymes,

7.1 Components of Biosensors

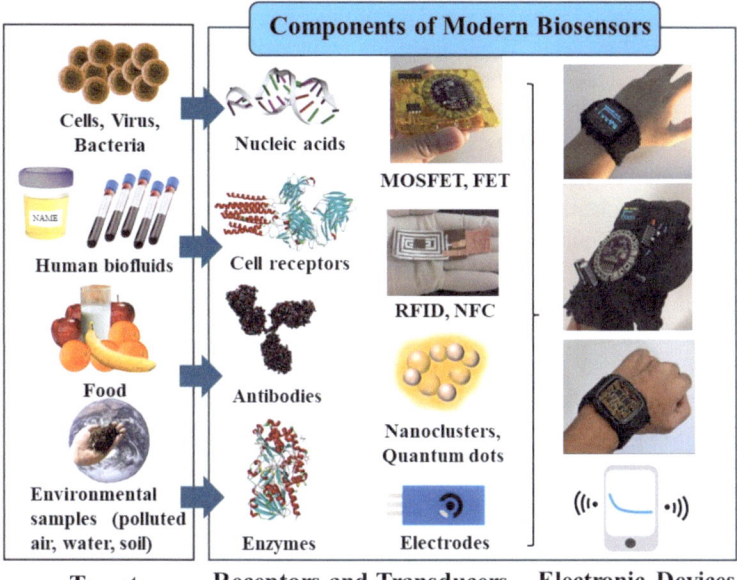

Fig. 7.1 Typical biosensor modules and their component elements. Reproduced with a permission from Refs. [8, 9]. Copyright 2024 by the Author and Royal Society of Chemistry and Copyright 2022 by the Author and American Chemical Society

peptides, oligomers, antibodies, and synthetic nanomaterials are examples of recognition components frequently utilized in biosensors. Enzymes are the most common type of them. For biorecognition to properly utilize the unique interaction, the sensor's surface pattern must also prevent any non-specific touch [11]. Many studies have focused on identifying surface modifications in biological fluids that have specific interaction capacities over long times. Lack of nanomaterials or material topologies that permit high enough sensitivity and distinct detection of the relevant analyte in the desired biological event has hampered the development of electrochemical biosensors. Important groups of biosensors, including immunosensors, respond differently depending on the pH, ionic strength, and unknown interfering species of the biofluids, all of which might change significantly. As a result, there has been a new focus on using nanotechnology to discover methods of using several enzymatic or artificial nanomaterial labels, for instance, to boost the signal for each event and decrease the size of an electrochemical sensor for point-of-need applications. To identify biological processes that take place at the electrode or probe contact, the biosensing design should also raise the signal-to-noise ratio more. These remaining problems may be resolved by a new generation of highly specialized, sensitive, selective, and reliable micro-to-nano chemical sensors and sensor arrays that combined knowledge in data processing, integrated circuit silicon technology, solid-state and surface physics, bioengineering, and data processing [12]. The essential transducing component in biosensing is an important physicochemical and electrochemical component that can stabilize signal data. It monitors and amplifies electrical properties from extracted data in biological systems. While there are many different recognition components in biosensing devices, electrochemical detection methods often include enzymes. Their distinct binding properties and biocatalytic activity are mostly to blame for this. Additional components of biorecognition include CRISPR-Cas, nucleic acids, aptamers, bacteria, cells, and antibodies. An immunosensor tracks binding activities during bio-electrochemical processes using antigens, aptamers, or antibodies. In electrochemistry, one of three things would typically happen to the reaction under study. First, the conductivity or conductometric in the medium clearly changes. Second, an amperometric or detectable high current response is present. Thirdly, there is a notable shift in the potentiometric potential or electromotive force. Other electrochemical detection methods, such as the field-effect transistor, which measures current using transistor technology due to a potentiometric effect at a gate electrode, and the impedimetric, which measures impedance for both resistance and reactance, have also significantly advanced. A variety of instruments that use these measuring techniques in their various versions have also surfaced [10].

For electrochemical biosensors to function and respond to biorecognition, the electrodes themselves are essential since biological reactions often take place on top of the electrode surface. Depending on the biosensing function chosen, the sticky sensing material, surface integration, or electrode size all significantly affect the electrode's object detection capability. Three electrodes are often needed for electrochemical sensing: a working electrode (also known as the sensing or redox electrode), a counter or auxiliary electrode, and a reference electrode. The reference electrode, which is often composed of silver or silver chloride, is placed far away from the reaction site in order to provide a consistent and known voltage or potential. The counter electrode connects to the electrolytic solution to enable a closed loop current in the sensor, while the working or sensing electrode functions as the transduction element in the biological response. Both chemical stability and conductivity are required for these electrodes. Therefore, depending on the analyte, platinum, gold, carbon paste, graphite, and silicon compounds are frequently used. The various detection limits and a concentration range of detection are excellent explanations of the analyte detection utilizing a specified electrode. The combination of nanotechnology and bioelectronics has produced innovative methods for optimizing and reducing the size of current microscale devices. With the use of several electrochemical transducer types, accurate measurement of electrical properties is becoming possible. When created and found nanomaterials have a higher surface-to-volume ratio, their electrical characteristics become more resistant to external stimuli, particularly when these structures continue to contract toward the atomic limit [13–15].

7.2 Detection of Small Molecules and Biochemical Conditions

Temperature, pH, ionic strength, and a host of other co-factors all affect most chemical reactions and biological activities. Thus, precise assessment of these attributes is essential in most research and technological domains as well as medicine. DNA has shown to be a popular material for the development of sensors that can evaluate these environmental variables. One explanation for this is the dynamic structure of the DNA molecule. The two strands of normal double-stranded DNA are joined by hydrogen bonds formed between the bases of the nucleotides. Depending on the base composition and surrounding conditions, the DNA duplex can adopt a variety of conformations, including the usual right-handed B-DNA conformation and the left-handed Z-DNA conformation. Furthermore, there are various ways in which DNA bases can stack to create triplex or quadruplex structures. Alterations in the surrounding aqueous solution, particularly those related to temperature, pH, or ionic strength, may promote conformational changes in DNA or the breaking of hydrogen bonds that split DNA strands. By introducing different changes, including fluorophore-quencher pairs, it has been possible to demonstrate the functionality of a variety of DNA sensors for monitoring different chemical or physical characteristics of the environment [16]. The melting temperature (T_m), or the point at which half of the molecules have melted and formed two DNA strands, varies for each duplex DNA structure in a given solution [17]. Thermometers have been made using this characteristic. The percentage of melted DNA molecules reflects a temperature-dependent balance between the two strands' re-annealing (hybridization) and melting of DNA. The number of hydrogen bonds and, subsequently, the quantity and composition of base pairs determine T_m for a given DNA molecule. Thus, a temperature sensor is inherited by the kinetics of hybridization of DNA molecules. Periodically, tiny "DNA-thermometers" have been made by taking use of the kinetics of hybridization between DNA molecules. DNA thermometers have been based on "molecular beacons," or DNA molecules that can create DNA hairpin structures. When complementary sequences at the ends of single-stranded DNA are employed to create DNA hairpins, intra-strand hybridization between the sequences can result in the formation of the hairpins [18].

The duplex region created by hybridization forms the stem of a DNA hairpin, while the nucleotides in between make up the loop. At a specific temperature, an equilibrium will form between single-stranded DNA molecules and closed hairpins. Below the melting point (T_m) of each individual DNA molecule, many molecules will form hairpins; however, when the temperature rises over T_m, the hairpins melt, and more molecules become single-stranded. The length and sequential sequence of the stem have an impact on the hairpin's T_m. Furthermore, the environment's composition—including pH and ionic strengths—will affect the T_m, therefore consideration of these factors is necessary when utilizing DNA-based sensors to measure temperature. The hybridization kinetics between a fluorophore and a quencher, which are located at the 3′ and 5′ ends of the DNA hairpin, respectively, are monitored by the molecular beacon using FRET. Low fluorescence is the outcome of the hairpin formation keeping the fluorophore near the quencher in the hairpin stem. Higher temperatures, however, cause DNA molecules to become single-stranded and detach the quencher from the fluorophore, which raises fluorescence levels. A sigmoid melting curve is the outcome of the fluorescent signal growing as the temperature rises [19–21].

7.3 Sensing Materials for Advanced Biosensors

With the aid of an engineering design orientated toward decentralized healthcare, the biosensors should be equipped with electrical connections, human–machine interface electrodes, and sensing materials (shown in Fig. 7.2) with suitable transducers for detecting different analyte targets. Piezoelectric material, for example, can be easily included into telemedicine wearables to measure applied pressures, body movement events, tissue tension, and strain. This substance transforms twists and deformation into electrical potential. Wearable healthcare sensors are essentially made of piezoelectric materials, including semiconductor materials, piezopolymers, piezoceramic, and piezoelectric green materials. Piezopolymers include, for example, poly(vinylidene fluoride-co-trifluoro ethylene) (PVDF-TrFE), poly(3,4-ethylenendioxythiophene)-poly(styrenesulfonate), and other materials. Even though some of them have low piezoelectric coefficients, their flexibility and light weight make them appropriate for biosensing applications [5, 22].

Stretchability and piezo materials' ability to heal themselves play a big part in prolonging wearable device life for continuous, uninterrupted monitoring in certain harsh scenarios, as after mechanical damage. Since the piezoelectric effect relied on the property that charge buildup is formed in an opposite direction on both sides of the dielectric materials, the sensors were largely utilized for physical sensing, pressure sensing, and movement disorders. For instance, to get around a constraint with piezoelectric inorganics intended for physical sensing, modified piezoelectric polymers were developed and used for neural interface technologies with deeper level sensing of neurotransmitters in the brain. Piezoresistive and capacitive movement sensors for health monitoring still have signal crosstalk, which results in inaccurate measurement, even though the creation of wearable sensors necessitates a large number of piezoelectric sensors [5, 7]. Another crucial concern with piezoelectric wearable sensors is their biocompatibility. For example, even though a lead-based piezoelectric sensor has a robust electromechanical response, the problem of lead contamination still needs to be addressed. Biomedical constraints necessitate new findings for lead-free piezoelectric materials. Biomaterials such as cellulose, chitosan, collagen, and amino acids were employed as biodegradable and biocompatible sources with a high degree of molecular sensing. These sources provided complementary elements to enhance sensing performance and were integrated with enzymatic oxidases to detect diabetes and other non-communicable diseases. Derived bio-responsive inks consisting of plasticizing glycerol, regenerated silk fibroin, and

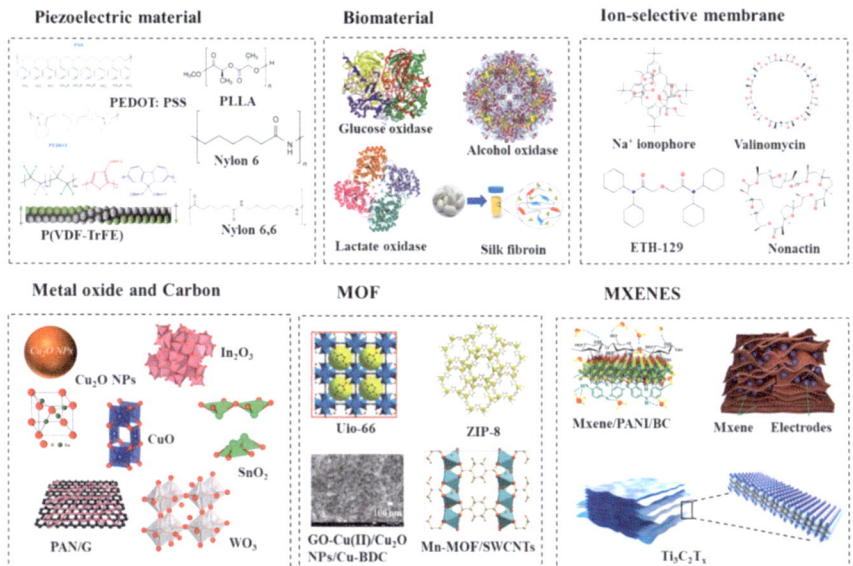

Fig. 7.2 Classes of sensing materials for sophisticated biosensors. Reproduced with a permission from Ref. [5]. Copyright 2023 by the Author and American Chemical Society

sodium alginate allowed for colorimetric pH monitoring, providing a viable route to the mass production of wearable interfaces for distributed sensing. One way to boost sensitivity is to combine polymeric additives with enzyme material to increase signal stability. For example, the salivary glucose-specific poly (MPC-co-EHMA-co-MBP) (PMEHB)-based telemetry mouthguard sensor was developed to immobilize the oxidase and track salivary glucose. This mount guard sensor additionally has an interference reject membrane made of cellulose acetate to help overcome the issue of interfering species and deliver a more accurate measurement of glucose. These biomaterials are sustainable, however because enzymatic materials frequently have a single useful life, they cannot be recycled [5, 9, 23–25].

Furthermore, biomaterials pose a problem due to their instability and rapid denature at room temperature, including aptamers, DNA, RNA, and antibodies. It could have led to a decrease in the performance of the worn sensors Therefore, it might not be suitable for many situations where there is intense activity or for a continual long-term health assessment. Using an ion-selective membrane as a sensing material to function under an opened circuit condition instead of an electronic component to realize an electrochemical redox reaction is one way to get around a barrier to constructing a compact wearable device. The conformal ion-selective electrodes can potentially have a smoothing function due to the integration of ion-to-electron transducers and plasticized polymeric membranes. In sweat sensing technology, ionophores that capture Na^+, Ca^{2+}, K^+, and NH^{4+} have proven to be effective thus far. This potentiometric sensor helped ion-selective electrodes gain important validation, opening the door for the creation of personalized recovery and hydration regimens for enhanced sports performance in the next wave of digital healthcare. From the standpoint of smart materials, ionophores have shown effective in sensing alkaline ions (Li^+, K^+, Na^+) and biologically significant ions (Ca^{2+}, Mg^{2+}, NO_3^-), however membranes for polyion selection have not advanced much. This was attributed to insufficient polyion recognition by the ligand selectivity. Still, several potentiometric sensors must deal with possible drifts caused by temperature changes in a sample or analyte. One idea to lessen this unwanted signal fluctuation is to use symmetric solid-contact potentiometric electrodes. This issue was resolved by the introduction of an interlayer of acrylate membrane, hydrophobic redox buffers, and enhanced ion-to-electron transducers, which helped to optimize signal sensing and lessen the need for frequent calibration in the development of high-performance wearable potentiometric sensors [5, 26, 27]. Certain metal oxides and active carbon have been carefully coupled with other novel active materials to improve the sensing capability toward certain biochemical targets or function as efficient ion-to-electron transducers for signal readout amplification. Numerous nanostructures, such as nanoparticles (NPs), nanowires (NWs), nanoplatelets, nanocubes, nanoflowers (NFs), and nanorods (NRs), have been employed for CuO-integrated sensors. These sensors are frequently able to identify toxic compounds that are harmful to human health. Flexible transparent pressure sensors with pulse detection could be made by floating a layer of graphene reinforced with polyacrylonitrile (PAN) freely on water or in the air. On the other hand, these pressure sensors showed excellent sensitivity at low operating voltage and exceptional loading cycle endurance. Without depending on enzymes that are prone to biodegradation, Cu(I) ions and poly(acrylic acid) gel-Cu(II) were able to detect creatinine and facilitate the identification of advanced kidney dysfunctions. Because this sensing material used irreversible detection, it could only be used once. Higher selectivity against other interferents requires more complex production, and some metal oxides may not be biocompatible, rendering them unsuitable for long-term use. Moreover, it is uncertain if this kind of sensing material results in skin irritation or cytotoxicity. It's probable that in order to get around these problems, metal oxides were discovered inside of microfluidic channels by attaching microfluidic channels to them [28]. To create a porous architecture with multi-biochemical activity, metal–organic frameworks (MOFs) were designed as porous coordination polymers (PCPs) paired with various biochemical transducers. PCPs have significantly advanced the disciplines of biosensing, renewable energy, gas storage, and catalysis due to their remarkable biocompatibility. MOFs are highly adaptable in terms of catalytic design and can host a range of guests, such as metal and semiconductor oxide NPs. These porous crystalline materials with infinite lattices are composed of flexible secondary building blocks that can be modified by organic linkers and metal cations. For example, the ultrahigh surface areas of MOF bridges are incomparable to those of zeolites and active carbons. The selectivity was enhanced by using condensed bimetallic nanoparticles within the MOF framework, which facilitated an interaction with the target analytes. For example, the ternary hybrid material of GO-Cu(II)/Cu_2O NPs/Cu-BDC was bonded with creatinine to assess kidney function. Excellent sweat lactate and glucose detection was demonstrated

7.3 Sensing Materials for Advanced Biosensors

by combining flexible amino-modified functionalized graphene sheets with a 2D orientated structure of $Cu_3(btc)_2$ nanocubes. Because heavy metals have such detrimental impacts on human systems, lead ion (Pb^{2+}) is one of the most hazardous and inert metals. Human serum samples were examined using the Mn-MOF/SWCNTs. With ZIF-8, a class of ZIFs consisting of zinc ions and 2-methylimidazolate, this intriguing material effectively recognized dopamine, which is associated with the stimulating functions of the central nervous system. Deficits in dopamine may be a factor in depression and other mental health issues. This material has generated a lot of interest because of its strong modifiability, affinity for analytes, and flexibility, especially given the increased MOF innovation prospects. Furthermore, MOF multi-metal sites might help catch analytes in the process of diffusing. Wearable sensors are unable to detect analytes for more than one use due to a nonreversible electrocatalytic process, despite the exceptionally high surface areas of MOFs. Future research should focus mostly on developing low-cost, versatile, and environmentally benign electroactive MOFs. Because of issues regarding biocompatibility, it is necessary to design novel ligands or coordinate existing ligands that mimic the metabiological activities provided by nature to fulfill the real demand for wearable sensing [7, 29].

To develop wearable sensors in storage conversion, harvesting, and sensing, MXenes—two-dimensional (2D) transition metal carbides, nitrides, and carbonitrides—are garnering a lot of attention. Mxenes' electrical and chemical properties can be effectively adjusted to increase the utility of hybrid sensing devices. Conventional metal-oxide semiconductors are commonly used for hazardous gas detection; however, their primary drawback is that their activation temperature is high. The high metallic conductivity and abundance of terminal functional groups in MXenes make them highly promising as gas, optical, humidity, strain/stress, and electrochemical sensors. $Ti_3C_2T_x$ Mxenes/PANI/bacterial cellulose electronic skin, for example, made it possible to monitor user motions as well as dangerously high levels of NH_3 gas. For this risky gas monitoring, the deep mechanism of PANI's protonation and deprotonation with both bases and emeraldine was employed. Self-powered devices have recently been developed to provide wearable sensors with consistent power supply while simultaneously prolonging their operational lifetime. Using Mxene-based micro supercapacitors and lithium-ion micro batteries, all self-powered integrated systems were developed as multifunctional inks that offered a consistent areal energy density. This integrated system responded to body motion very quickly and was quite responsive. Even though 2D Mxenes have recently improved, their cytotoxicity is still a major problem that needs to be resolved before these materials are used commercially. Moreover, it is still hard to fabricate Mxene nanosheets with the right size, dopants, surface terminations, and arbitrary spatial organization, which prevents Mxenes from being used in materials for sensing under controlled conditions. These materials have demonstrated promising performance for sensing applications, enabling the next generation of wearable telemedicine sensors to perform enthusiastically [5, 30, 31].

7.4 Kidney-Disease and Global Public Health Issues

Globally, chronic kidney disease (CKD) ranks 14th in terms of the greatest cause of years lost from life. It is estimated that 850 million people worldwide suffer from renal disease, with the majority residing in low-income and lower-middle-income nations. The World Health Organization (WHO) lists heart disease, cerebrovascular disease, coronary artery disease, cancer, diabetes, and chronic cardiovascular disease as the primary non-communicable diseases (NCDs) that result in early mortality and disability. Kidney disease, either acute kidney injury (AKI) or chronic kidney disease (CKD), is noticeably missing from this list. It should be noted that CKD is more common among people with AKI, and vice versa [32]. It is crucial for healthcare that the death rate from non-communicable diseases (NCDs) be reduced by one-third by 2030. CKD, which affects 8–16% of the global population, is defined as a persistent impairment in kidney structure or function (e.g., glomerular filtration rate [GFR] < 60 mL/min/1.73 m^2 or albuminuria $\geq$ 30 mg per 24 h) that persists for more than three months. Chronic kidney disease (CKD) is a chronic condition characterized by long-term changes in kidney structure, function, or both, with potential health consequences. Imaging-detectable structural abnormalities include tumors, cysts, malformations, and atrophy. On the other hand, renal dysfunction might present as elevated serum levels of creatinine, cystatin C, or blood urea nitrogen, as well as development delay in children, edema, changes in urine production or quality, and hypertension. Renal fibrosis, in whatever form, is the most prevalent pathological manifestation of chronic kidney disease (CKD), independent of the underlying insult or illness [33]. The most well-known factors influencing the incidence of chronic kidney disease (CKD), particularly in areas with developed economies, are population increase, aging, and the rising prevalence of diabetes, heart disease, and hypertension. Normally, in high-income nations, CKD affects up to 33% of diabetics and 20% of people with hypertension. It has been determined that controlling diabetes and cardiovascular disease will lessen the growing burden of CKD. It's interesting to note that men are more likely to have end-stage renal disease even though women have a larger frequency of CKD than do men. Diabetes mellitus and hypertension are the most prevalent underlying illnesses linked to chronic kidney disease (CKD), especially in high- and middle-income nations [34].

7.5 Wearable Sensors for Kidney Monitoring

Several cutting-edge telemedicine solutions have been unveiled to create a new avenue for the medical industry. These recently created instruments make it easy to monitor and interact with patients from a distance. Chemical transducers with wearable electronics and the Internet of Things (IoT) integrated may be utilized with telemedicine technologies to build noninvasive smart sensors. Overuse of painkillers, high intake of salt and/or sugar, and excessive alcohol use are common factors that

7.5 Wearable Sensors for Kidney Monitoring

raise the risk of renal dysfunction in a fast-paced, hectic lifestyle. Since chronic kidney disease affects 8–16% of the world's population and is predicted to rank as the fifth leading cause of death by 2040. For example, advanced kidney disease, cognitive decline, osteoporosis, or even diabetes are examples of complex health conditions. Serum creatinine levels are a recommended indicator for assessing kidney health since they show how efficiently the kidneys can filter excessive metal ions and uremic poisons from the body. Serum creatinine levels more than 1000 μM, or 11.31 mg/dL, indicate renal failure; for healthy kidneys, the range is 45–140 μM, or 0.51–1.58 mg/dL. Serum creatinine levels measured by venous blood punctures might cause bruises, infections, stress, or clinic avoidance because of repeated blood tests; also, it can take several hours to receive the lab's results. A fiber-based sensing device that is selective for tear creatinine, known as lab-on-a-chip for kidney monitoring (given in Fig. 7.3), was created by utilizing a copper-containing benzene dicarboxylate (BDC) metal–organic framework (MOF) that was bonded with graphene oxide-Cu(II) and hybridized with Cu_2O nanoparticles (NPs) [35].

Low-cost, sustainable, and environmentally friendly cotton fiber was utilized in place of the screen-printed carbon electrode (SPCE) as a textile-based electrode to create an economical kidney monitoring device. A three-electrode system connected to an LED monitoring screen and microcontroller unit (MCU) is shown in Fig. 7.4 as a circuit design. A microcontroller unit (MCU) and an LED monitoring panel were electrically connected to the three-electrode system. The as-prepared hybrid textile working (WE), reference (RE), and counter electrodes (CEs) are also depicted in the picture [35]. Utilizing the basis set of def2-SVP and the computational density functional theory (DFT) approach with CAM-B3LYP, the interaction between creatinine and Cu^+-Cu-BDC MOF of two distinct BDC chains was evaluated. Through Cu–O bonding, the cuprous ions of the oxidized Cu_2O NPs and Cu-BDC MOF were linked to create the Cu^+-Cu-BDC MOF. An important characteristic of Cu-BDC MOF is that it affects binding energies via three distinct =O, –N, and $–NH_2$ sites with the 2-amino-1-methyl-5H-imidazole-4-one structure of creatinine. As more cuprous

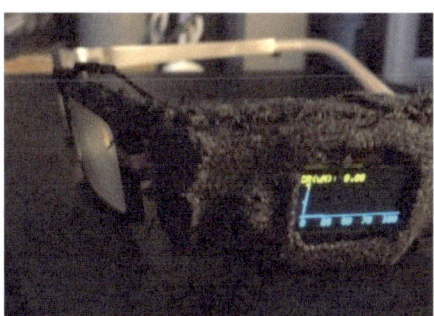
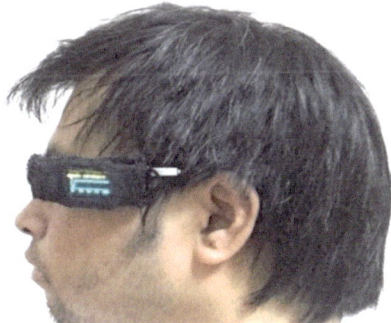

Fig. 7.3 Complete integrated lab for kidney monitoring on eyewear. Reproduced with a permission from Ref. [35]. Copyright 2021 by the Author and American Chemical Society

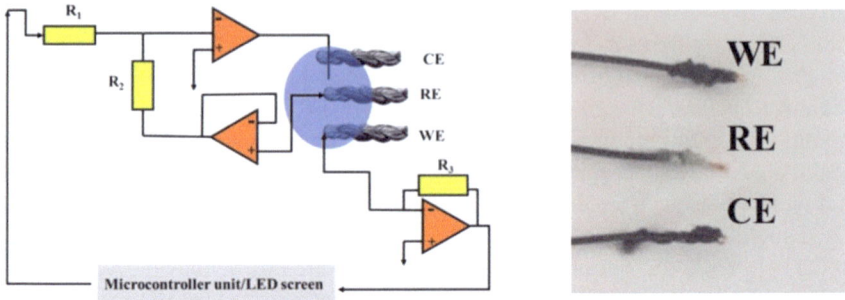

Fig. 7.4 The textile-based sensor's electrochemical circuit design is combined with a microcontroller for real-time monitoring and works with ready-made textile working, reference, and counter electrodes. Reproduced with a permission from Ref. [35]. Copyright 2021 by the Author and American Chemical Society

ions formed per shorter BDC segments, creatinine's preference toward the uncoordinated copper clusters increased. For the heterogeneous reaction, the binding sites of cuprous ions with creatinine that were most preferred were at $-NH_2$ [13, 14, 35]. The procedure of fabricating the carbon black/Cu-BDC MOF/GO-Cu(II)/Cu_2O NPs based on cotton fiber and wrapped around an electrode wire to create a robust and environmentally resistant working electrode is schematically depicted in Fig. 7.5a. The SEM picture of the hybrid materials utilized to coat a single functional electrode for single use is displayed in Fig. 7.5b. Observed was the uneven ternary complex coating that persisted on the cotton fiber's intertubular regions. After completing many redox cycles for the hybridization, the as-prepared Cu-BDC MOF/GO-Cu(II)/

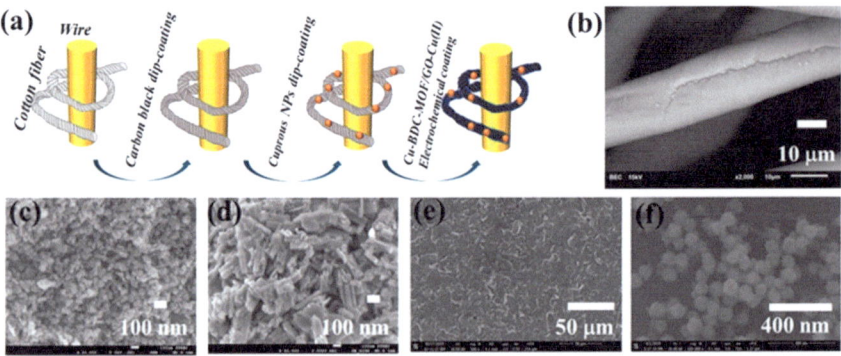

Fig. 7.5 a The method of fabrication used to create carbon black based on cotton fiber, Cu-BDC MOF, GO-Cu(II), and Cu_2O nanoparticles twisted on an electrode wire for detecting tear-creatinine. Enhanced in SEM images of **b** carbon black derived from cotton fiber/Cu-BDC MOF/GO-Cu(II)/Cu_2O NPs, **c** GO-Cu(II)/Cu-BDC MOF/Cu_2O NPs on SPCE, **d** GO-Cu(II) on SPCE, **e** Cu-BDC MOF on SPCE, **f** and Cu_2O NPs on SPCE. Reproduced with a permission from Ref. [35]. Copyright 2021 by the Author and American Chemical Society

7.5 Wearable Sensors for Kidney Monitoring

Cu_2O NPs on SPCE displayed well-dispersed and nano-sized features of pores with the higher surface as well as flaky structures, as illustrated in Fig. 7.5c. The target molecule, creatinine, was well-entrapped by the ternary hybrids' porous structures. The exfoliation of hybridized graphene sheets and the high specific surface of generated pore structures were responsible for this special quality. The layered Cu-BDC MOF's cubic structure is shown in Fig. 7.5d. The flaky microstructure of graphene oxide combined with cuprite ions is seen in Fig. 7.5e. Its multilayer construction allowed for the observation of agglomerated sheets. While R_{CT} deduced the reaction kinetics rate of these hybrid materials, Electrochemical Impedance Spectroscopy (EIS) was used to confirm the charge transfer resistance of the tear-based creatinine sensor. It appears that many hundred ohms of R_{CT} indicate very sluggish reaction kinetics at the heterogeneous electrode surface, while low R_{CT} suggests very quick response kinetics of delocalized cuprous and cuprite ions.

As shown in Fig. 7.6a, b, respectively, the Nyquist plots were obtained for the impedance spectroscopy of Cu-BDC-MOF, GO-Cu(II), Cu_2O NPs, and Cu-BDC MOF/GO-Cu(II)/Cu_2O NPs generated at the first and the tenth CV scans using 1 mM $[Fe(CN)_6]^{3-}/^{4-}$. By extending the semicircle diameters to the Randles equivalent circuit, the R_{CT} was calculated. By taking the intersection of the Nyquist semicircle with the Z_{REAL} axis at the high-frequency region, the electrolyte resistance, or RE, was determined. The Warburg area of 45° constant phase element (CPE) in the intermediate frequency of the Nyquist plots was used to investigate electrolyte ion diffusion close to the electrode interface. Ordered by Cu_2O NPs < GO-Cu(II)/Cu-BDC MOF/GO-Cu(II/Cu_2O NPs < Cu-BDC MOF) were the Warburg lengths. In

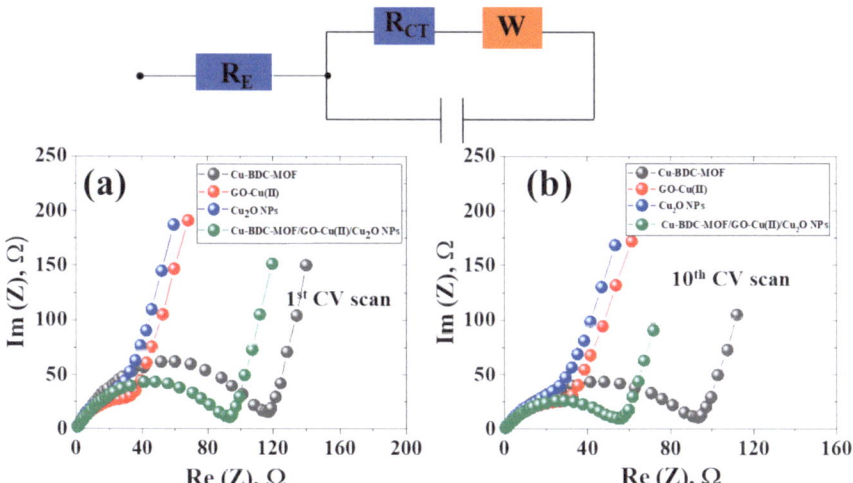

Fig. 7.6 Nyquist plots were produced on SPCE in the frequency range of 0.1–100,000 Hz for Cu-BDC MOF, GO-Cu(II), Cu_2O NPs, and patulous Cu-BDC-MOF/GO-Cu(II)/Cu_2O NPs hybrids **a** at the first CV scan and **b** at the tenth CV scan. Reproduced with a permission from Ref. [35]. Copyright 2021 by the Author and American Chemical Society

comparison to Cu-BDC MOF, it showed that the ternary complex offered quicker ion kinetics. Figure 7.7a–f display the empirical machine-learning fitting curves for the relationships between serum and tear creatinine that were derived using various expected approaches of neural networks, linear regression, closest neighbors, random forests, decision trees, and Gaussian processes. Supervised clinical data from renal patients was fed into machine learning algorithms.

However, because the fitting line divided into several unique regimes and was independent of serum creatinine in some concentration ranges, the decision tree, closest neighbor, random forest, and Gaussian process approaches were not effective for the measurement of continuous tear creatinine variables. Thus, the wearable sensor selective for tear creatinine might use the neural network and linear regression techniques as its primary machine learning-based processing units. Figure 7.7g showed the confusion matrix used to estimate the relationship between tear and serum creatinine. A confusion matrix, which represents the real blood creatinine contents to

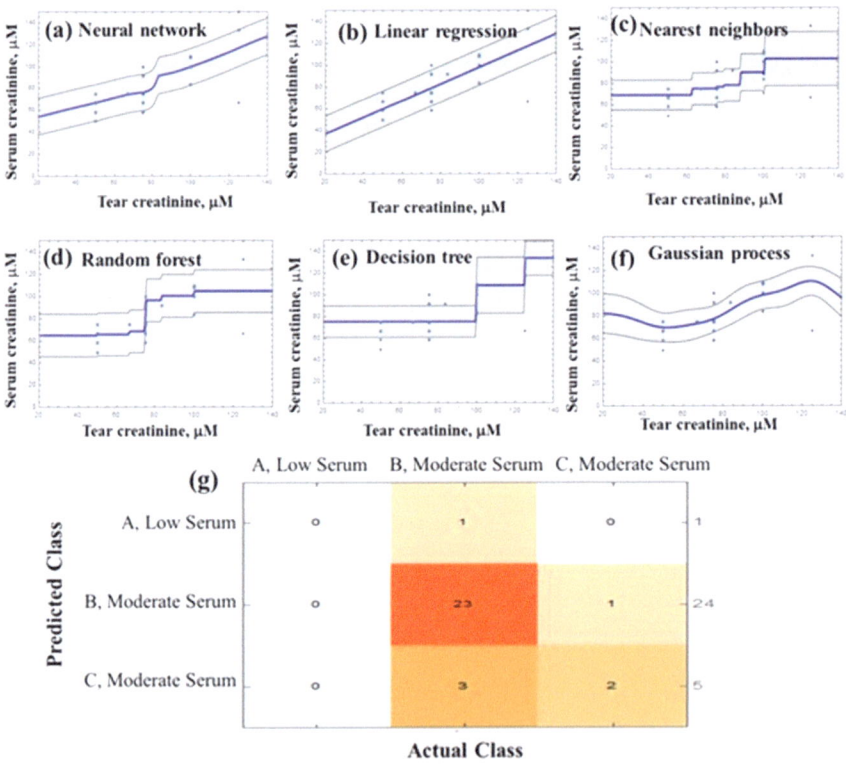

Fig. 7.7 Using techniques of **a** neural network, **b** linear regression, **c** closest neighbors, **d** random forest, **e** decision tree and **f** Gaussian process, machine learning-fitting data sets of serum and tear creatinine relations are performed. An illustration of a confusion matrix (**g**) utilizing tear creatinine to show the actual and expected classes for serum creatinine. Reproduced with a permission from Ref. [35]. Copyright 2021 by the Author and American Chemical Society

assigned serum creatinine classes, summarizes the outcome of the machine learning-based classifier with an algorithm accuracy of 83.34%. In the end, the lab-on-a-chip for kidney monitoring addressed a critical need to develop a wearable sensor guided by machine learning and integrated with disposable hybrid textile electrodes for kidney monitoring. This is especially important for vulnerable populations during unforeseen disruptive pandemics, as it reduces the risk of infection [5].

References

1. J. Goode, J. Rushworth, P. Millner, Biosensor regeneration: a review of common techniques and outcomes. Langmuir **31**(23), 6267–6276 (2015)
2. N. Granqvist et al., Characterizing ultrathin and thick organic layers by surface plasmon resonance three-wavelength and waveguide mode analysis. Langmuir **29**(27), 8561–8571 (2013)
3. A.R. Ferhan, J.A. Jackman, N.-J. Cho, Integration of quartz crystal microbalance-dissipation and reflection-mode localized surface plasmon resonance sensors for biomacromolecular interaction analysis. Anal. Chem. **88**(24), 12524–12531 (2016)
4. A. Kueng, C. Kranz, B. Mizaikoff, Scanning probe microscopy with integrated biosensors. Sens. Lett. **1**(1), 2–15 (2003)
5. S. Kalasin, W. Surareungchai, Challenges of emerging wearable sensors for remote monitoring toward telemedicine healthcare. Anal. Chem. **95**(3), 1773–1784 (2023)
6. A.J. Bandodkar, I. Jeerapan, J. Wang, Wearable chemical sensors: present challenges and future prospects. ACS Sens. **1**(5), 464–482 (2016)
7. J. Heikenfeld et al., Wearable sensors: modalities, challenges, and prospects. Lab Chip **18**(2), 217–248 (2018)
8. S. Kalasin, W. Surareungchai, Artificial intelligence-aiding lab-on-a-chip workforce designed oral [3.1.0] bi and [4.2.0] tricyclic catalytic interceptors inhibiting multiple SARS-CoV-2 protomers assisted by double-shell deep learning. RSC Adv. **14**(37), 26897–26910 (2024)
9. S. Kalasin, P. Sangnuang, W. Surareungchai, Intelligent wearable sensors interconnected with advanced wound dressing bandages for contactless chronic skin monitoring: artificial intelligence for predicting tissue regeneration. Anal. Chem. **94**(18), 6842–6852 (2022)
10. E.O. Polat et al., Transducer technologies for biosensors and their wearable applications. Biosensors **12**(6), 385 (2022)
11. C. Karunakaran, T. Madasamy, N.K. Sethy, Enzymatic biosensors, in *Biosensors and Bioelectronics* (Elsevier, 2015), pp. 133–204
12. J. Baranwal et al., Electrochemical sensors and their applications: a review. Chemosensors **10**(9), 363 (2022)
13. S. Kalasin et al., Salivary creatinine detection using Cu(I)/Cu(II) catalyst layer of supercapacitive hybrid sensor: wireless IoT device to monitor kidney diseases for remote medical mobility. ACS Biomater. Sci. Eng. (2020)
14. S. Kalasin et al., Evidence of Cu(I) coupling with creatinine using cuprous nanoparticles encapsulated with polyacrylic acid gel-Cu(II) in facilitating the determination of advanced kidney dysfunctions. ACS Biomater. Sci. Eng. **6**(2), 1247–1258 (2020)
15. T. Lim, H. Zhang, S. Lee, Gold and silver nanocomposite-based biostable and biocompatible electronic textile for wearable electromyographic biosensors. APL Mater. **9**(9) (2021)
16. Y. Li et al., Hierarchical assembly of super-DNA origami based on a flexible and covalent-bound branched DNA structure. J. Am. Chem. Soc. **143**(47), 19893–19900 (2021)
17. P. Gao et al., A novel assay based on DNA melting temperature for multiplexed identification of SARS-CoV-2 and influenza A/B viruses. Front. Microbiol. **14**, 1249085 (2023)

18. P. Shah et al., Noncanonical head-to-head hairpin DNA dimerization is essential for the synthesis of orange emissive silver nanoclusters. ACS Nano **14**(7), 8697–8706 (2020)
19. C. Fang, Y. Huang, Y. Zhao, Review of FRET biosensing and its application in biomolecular detection. Am. J. Transl. Res. **15**(2), 694 (2023)
20. M. Imani et al., Recent advances in FRET-based biosensors for biomedical applications. Anal. Biochem. **630**, 114323 (2021)
21. C. Haasnoot et al., Structure, kinetics and thermodynamics of DNA hairpin fragments in solution. J. Biomol. Struct. Dyn. **1**(1), 115–129 (1983)
22. S. Kalasin, P. Sangnuang, W. Surareungchai, Satellite-based sensor for environmental heat-stress sweat creatinine monitoring: the remote artificial intelligence-assisted epidermal wearable sensing for health evaluation. ACS Biomater. Sci. Eng. (2020)
23. J.R. Sempionatto et al., Wearable chemical sensors for biomarker discovery in the omics era. Nat. Rev. Chem. **6**(12), 899–915 (2022)
24. N. Promphet et al., Non-invasive wearable chemical sensors in real-life applications. Anal. Chim. Acta **1179**, 338643 (2021)
25. G.N. Islam, A. Ali, S. Collie, Textile sensors for wearable applications: a comprehensive review. Cellulose **27**, 6103–6131 (2020)
26. H. Zhao et al., Recent advances in flexible and wearable sensors for monitoring chemical molecules. Nanoscale **14**(5), 1653–1669 (2022)
27. E. Zdrachek, E. Bakker, Potentiometric sensing. Anal. Chem. **93**(1), 72–102 (2020)
28. M.A.A. Mamun, M.R. Yuce, Recent progress in nanomaterial enabled chemical sensors for wearable environmental monitoring applications. Adv. Funct. Mater. **30**(51), 2005703 (2020)
29. H. Meskher, S.B. Belhaouari, F. Sharifianjazi, Mini review about metal organic framework (MOF)-based wearable sensors: challenges and prospects. Heliyon **9**(11) (2023)
30. J.F. Olorunyomi et al., Metal–organic frameworks for chemical sensing devices. Mater. Horiz. **8**(9), 2387–2419 (2021)
31. A. Sharma et al., Involvement of metal organic frameworks in wearable electrochemical sensor for efficient performance. Trends Environ. Anal. Chem. **38**, e00200 (2023)
32. A.C. Webster et al., Chronic kidney disease. Lancet **389**(10075), 1238–1252 (2017)
33. A. Francis et al., Chronic kidney disease and the global public health agenda: an international consensus. Nat. Rev. Nephrol. 1–13 (2024)
34. P. Romagnani et al., Chronic kidney disease. Nat. Rev. Dis. Primers **3**(1), 1–24 (2017)
35. S. Kalasin, P. Sangnuang, W. Surareungchai, Lab-on-eyeglasses to monitor kidneys and strengthen vulnerable populations in pandemics: machine learning in predicting serum creatinine using tear creatinine. Anal. Chem. **93**(30), 10661–10671 (2021)

Chapter 8
Self-sensing Intelligent Micro and Nanorobots for Monitoring Systems

Abstract Imagine a future in which robots assist our elderly grandparents, operate cars, and teach children. These futuristic possibilities first emerged ten years ago thanks to rapidly advancing robotics and artificial intelligence technologies. However, humans lack specially designed psychological frameworks to understand and interact with these advanced technologies. People rely on their ability to think socially, which has developed over thousands of years as a means of communicating with other people. Traditional medical examination methods have limitations. Biosensors with electrochemical techniques may identify biological information in samples. Strict sample conditions are typically required for traditional biosensors. As robotic research continues to advance into modern society, the development of micro and nanorobots has significantly expedited the field of precision medicine in modern biomedicine, particularly in the integration of monitoring systems. When combined, biomedicine and micro- and nanorobotics provide unique advantages and fascinating opportunities for illness early detection. Nanoscale robots have garnered a lot of interest due to their small size, flexible mobility, and promise for a range of in situ treatments. Nanoscale robots are perfect for nearly noninvasive or minimally invasive in-person diagnosis and treatment because of their small size and flexible movement, especially when it comes to targeted drug delivery for disease therapy in confined organ tissues.

8.1 Origin of Driven Micro and Nanorobots

Many species in nature have magnetoreception, which allows them to understand navigational cues from geomagnetic fields. Examples of these organisms include birds, bats, and butterflies, which enable them to fly or swim over long distances. Certain species, such as salmon and sea turtles, can detect the Earth's magnetic field and use this sense to navigate. These animals' magnetoreception aids in the formation of regional maps. Studies conducted on migrating birds demonstrate that they use the eye's cryptochrome protein, which relies on the quantum radical pair mechanism to sense magnetic fields. Furthermore, magnetotactic bacteria align themselves in accordance with the magnetic field lines to reach areas where the concentration of oxygen is best. Magnetotaxis is the biological phenomenon whereby microorganisms position themselves to move in response to the magnetic features of their surroundings. Similarly, the navigation of magnetic fields, primarily generated by charges moving through electric fields and intrinsic magnetic materials, allows nanoscale and microscale objects to move along a predetermined path, attracting significant attention due to its immense potential in environmental remediation and biomedicine. These tiny objects are commonly known as nanomotors, nanoswimmers, or magnetically propelled micro- and nanorobots. A robot must perform three functions. It should notice its surroundings, act automatically, and make decisions [1]. Micro and nanorobots are defined as diminutive machines that perform tasks, such as repairing at the cellular level or fabrication and monitoring at the molecular level. They operate on a micro- or nanoscale, powered either by their own chemical energy or external control, and can even be designed to carry out specific tasks in deep regions. The ideal micro- and nanorobots should be able to carry out duties by fabricating distinctive micro- and nanostructures, functional material decorating, or encapsulation and functionalization with diagnostic or therapeutic substances. Occasionally, they can perform delivery tasks, either by traveling in an experimentally and theoretically optimized course or by following a user-defined path to the desired locations. Subsequently, they carry out operations like eradicating cancerous tissues and sick cells, cleaning up the environment for regenerative medicine, or even sending data wirelessly to health monitoring systems. The inherent ability to locomotion is vital advantage for micro and nanorobots to carry out specific tasks in very small channels [2].

Chemical catalysis, such as the generation of gaseous oxygen and carbon dioxide, can dictate the transport of micro- and nanorobots to a target. The chemical reaction and external energy sources, including electromagnetic fields, acoustic waves, and sound, enable the motion of microbes, sperm, and pathogens. Chemical-powered micro and nanorobots outperform conventionally powered ones in speed, but they lack directional mobility. They also require fuels that are toxic and reactive, such as urea, sodium borohydride, hydrochloric acid, and hydrogen peroxide. On the other hand, although they move more slowly, micro and nanorobots driven by external physical forces, including light, magnetic, ultrasonic, and electric fields, do not require dangerous chemical fuels. Light-driven micro- and nanorobots have the potential to navigate across water, but their chemical composition may require H_2O_2 in addition to a powerful light source, potentially making them less biocompatible. However, there is currently limited biological usage and insufficient evidence to support it. Up till now, magnetically powered micro/nanomotors have been the most studied and used in a wide range of biomedical applications as well as environmental management and cleanup. They also address most of the drawbacks associated with alternative propulsion concepts. Magnetic resonance imaging (MRI) equipment can control clinical microrobots, allowing for the simultaneous use of medical MRI technology for microrobot imaging, target tracking, propulsion, and motion control. Clinical ultrasonography systems can navigate ultrasound trapping miniature robots, offering significant potential for vasculature investigation [3]. Importantly, there are apparent benefits to manipulating diminutive robots with a magnetic field through actuation systems. Intelligent robots have the ability to navigate into deep or otherwise inaccessible tissues within the body. Essentially, due to the contactless propagation of magnetic forces, magnetic fields provide a non-intrusive way to alter movement, allowing for remote maneuverability. Besides, such wireless actuation technology provides unrestricted motion for miniature machines while preserving the integrity of their local chemical and biological environment. Lastly, it runs without fuel or self-propelling. Relying on a magnetic field for propulsion is a potential strategy that does not require liquid fuel to drive microrobots. They also demonstrate notable movement speeds and even adopt distinct postures [4, 5].

8.2 Micro and Nanorobots Powered by Chemicals

The mechanism responsible for the directed force has been the subject of great debate ever since Au/Pt nanowires were shown to be capable of autonomous motion. Porous glass segments plated in platinum (Pt) on PDMS surfaces were used to steer the millimeter-scale objects. They attributed the motion of their microrobot to the recoil force of the oxygen bubbles created by the Pt catalyst, as seen in Fig. 8.1a. Ni/Au nanowires containing an Au end glued to a silicon wafer exhibited rotational motion upon the addition of hydrogen peroxide. A comparable method was suggested for these nanowires. The circular motion, according to the authors, was generated by the uneven development of oxygen. When Au/Pt nanowires were used to test this mechanism, it was discovered that the Pt end moved first and that the nanowires moved against expectations. The velocity of the nanowires could no longer be explained by the macroscopic bubbles. Based on the theory that the oxygen produced during the fuel's breakdown reduces the interfacial tension between the solution and the nanowire, a different interfacial tension model was put forth. Based on the Brownian ratchet mechanism, alternative model could be derived. It was stated that the viscosity close to the Pt decreased because of the oxygen generated at one end of the rod, enabling thermal motion to preferentially push it in that direction [6]. The Preferential reduction of hydrogen peroxide at the Au end and preferential oxidation of hydrogen peroxide at the Pt end are demonstrated by the bimetallic Au/Pt nanomotors in Fig. 8.1b. Due to an overabundance of hydrogen ion on the Pt side and an electric field directed from the Pt end to the Au end, the negatively charged rod advances along with the Pt side. The interior walls of the tubular structures discussed in the previous section are composed of active materials that facilitate the breakdown of chemicals into gas molecules. Although those materials can be catalytic or noncatalytic and can break down a range of fuels, hydrogen peroxide has been the main fuel that has been considered thus far. The motion of microjets is explained in three stages: The catalytic material is initially hydrated by the fuel solution, providing energetically favorable nucleation sites where oxygen accumulates and forms bubbles. In a later stage, usually the larger hole, the bubbles go toward one entrance of the microtube and are discharged, creating another motion step [7].

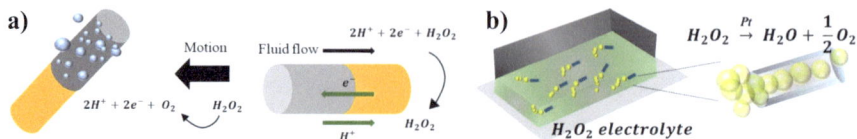

Fig. 8.1 a The processes for the motion of self-propelled nanowires are the electrokinetic mechanics of Pt/Au nanowires and the bubble propulsion of an Au/Ni nanowire. **b** Schematically, many microjets with propulsion via released bubbles are illustrated

8.3 Medical Micro and Nanorobots in Precision Medicine

As seen in Fig. 8.2, reducing the size of robotic platforms may enhance patient diagnosis and medical care for precision medicine. With these nanorobotic surgeons, we could perform a range of medical procedures on hard-to-reach parts of the body. Over the past decades, there has been progress in the creation of medical micro and nanorobots. Nevertheless, a main challenge realizing this research area is the need to establish these intelligent machines into more clinical advancement. Micro and nanorobots could benefit in medical prognosis by suggesting an accurate diagnosis of chronic diseases and even sending crucial signals through the isolation of microbes or the real-time assessment of tissue's physical conditions. The integration of micro and nanorobots with medical imaging functionalities would facilitate accurate internal positioning. Determining what constitutes a micro and nanorobot has demonstrated complexity due to the major differences in small-scale robotic systems from their large-scale counterparts. When a new area of research is developed inside an existing topic, it mostly involves in in vitro studies. In the early years of automobile use on city streets, they were called horseless carriages. The design of minuscule machines faces several hurdles when compared to existing automobile combustion engines, electrical motors, or fossil fuel power. When an object is smaller than a micron, its inertia is controlled by fluid viscosity and Reynolds number (Re) is negligible. Fluid variables such as Brownian motion which governs water molecules to collide randomly with microscopic objects. Nevertheless, Brownian dynamics can hinder the directionality of a motile micro and nano machines and must be taken into account when designing structures below microscale [8]. Ostensibly, locomotion of diminutive machines is dominated by Brownian motion and low Reynolds numbers at very small scale. The empirical formation of micro and nano machines is to create intelligent engines that can persistently move and exert enough thrusting force to overcome drag forces in the microscopic environment. Hence, what motivates the model and construction of minute robots is the need for highly synergistic and novel materials that can continuously transform different energy sources into propulsion. To produce directional motion, chemically propelled microrobots require an uneven dispersion of catalytic material; magnetically propelled micromotors employ magnetic materials to cause a micro-engineered structure to rotate; and ultrasonically propelled motors employ a structure with asymmetries in density to generate pressure gradients that facilitate their movement. The initial invention of micro and nano engines used in small scale

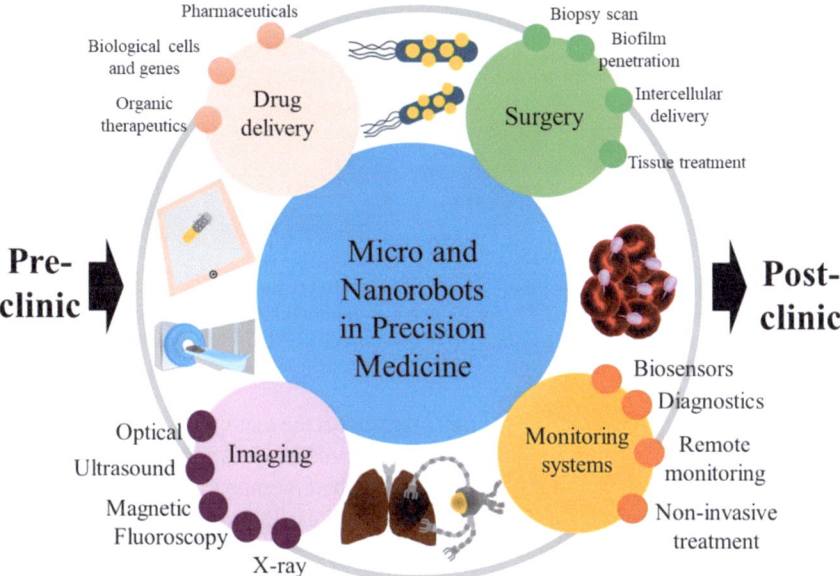

Fig. 8.2 Diagram shows the many applications of micro and nanorobotics in precision medicine today, such as medical imaging, delivery, surgery, diagnosis and monitoring systems

robots depended on trivial designs and fabrication methods. By electrochemically minimizing metallic salts within nano and micro symmetrical holes, these preliminary nanorobots were produced. The large-scale mass production with more than a million machines in each batch and the competence to introduce various electroactive materials, such as metals, polymers, semiconductors and different designs, including hollow tubes and porous wires in the same fabrication are the benefits of this well-researched invention technology. One more bottom-up approach is self-assembly. This includes the sequentially charged materials being assembled layer by layer, the self-organized polymers being created to construct catalytic materials-filled bowl-shaped somatotypes, and the colloidal materials being joined to form magnetic links and planned structures [9].

8.4 Targeted Drug Delivery

Micro and nanorobots that can be piloted into diseased tissue may serve as an adaptable platform for delivering a range of applications. Pharmaceuticals, biologics, live cell-based therapies, inorganic medicines, and monitoring systems are a few applicable examples as given in Fig. 8.3a. Likewise, the artificial stimuli that direct and steer diminutive machines can be entrapped to enhance drug targeting by stimulating the release of the therapeutic capsules when the diminutive machines reach a certain

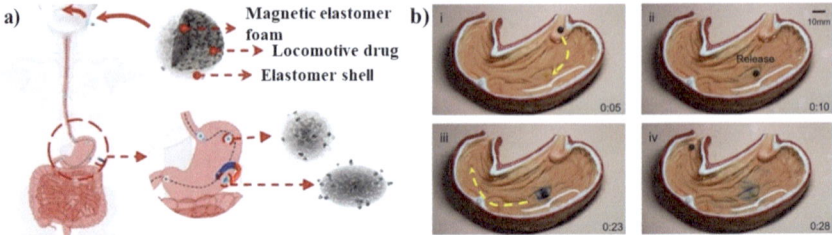

Fig. 8.3 **a** Platform of diminutive robots piloted into diseased tissues. **b** In vitro drug delivery of the hard-magnetic elastomer foam (HEF) capsule in a locomotion through the human stomach model. Reproduced with a permission from Ref. [10]. Copyright 2023 American Chemical Society

position in bodies. Many relevant factors often affect the therapeutic efficacy of medications on sickness. In traditional treatment procedures, it is normally necessary to provide high doses repeatedly to achieve the planned therapeutic impact, which may regrettably result in an increase in adverse side effects. Taking another approach, the diminutive machines can carry treatments by using a variety of mechanism in molecular binding, such as ionic, electrostatic and even covalent interactions to trap medicinal molecules on active materials. The therapeutic cargo, including drug molecules, stem cells, and genes are then discharged via a variety of methods, including trigger and release steps induced by applying external fields like, sound, near infrared and magnetic waves. Furthermore, autonomous release can be triggered by electrolyte condition changes, like pH or solution temperature. As shown in Fig. 8.3b, In vitro drug delivery of the hard-magnetic elastomer foam (HEF) capsule in a locomotion was demonstrated with the human stomach model. The controlled drug release rate ranging from 0.02 to 1.7 mL/min activated by the capsule was achieved. The HEF capsule enabled of functioning as a soft robot to perform magnetically driven drug delivery for biomedical applications in the human body [10].

The pharmaceutical business generates several synthetic medications designed to both cure and prevent illnesses. Unproductive pharmacokinetic properties of medications, such as short half-lives, poor biodistribution, and quick drug clearance in the body, have shown limitations in current healthcare. Hence, it frequently reduces the efficacy of therapeutic strategies. Nonetheless, conventional therapy also depends on high supplied doses and multiple high-dose injections are needed to achieve the intended therapeutic effect, which may lead to toxicity and other problems like rash, constipation, drowsiness or even thrombosis and inflammation [11]. Micro and nano machines and diminutive robots can provide a motile platform that can provide a precise dosage in the targeted place, hence solving the problem of relying on the systemic release of large therapeutic dosages. Pharmacological materials have also been directly captured via electrostatic interactions on the surface of micro- and nanorobots. According to reports, a negatively charged polypyrrole-polystyrene sulfonate section of an ultrasound-driven nanorobot was loaded with the positively charged brilliant green antibacterial drug by electrostatic forces. The electrostatic interaction was constant at pH 7. In a narrow channel of tissues inside body areas, untethered mobile microrobots have achieved in providing minimally invasive theragnostic operation. It is realized to employ pliable, biodegradable micro and nanorobot capsules with programmable near-infrared activation and magnet power. The micro machine consisted of doxorubicin-loaded photocleavable chains bound with chitosan. The pharmacokinetics of doxorubicin trigger and release were activated by employing a near-infrared field [12]. In addition, lysozyme, a naturally antimicrobial enzyme present in the innate human system, was exploited to compromise the micro and nano machines without providing any hazardous waste-products. A research progress claimed that a methacryloyl chloride and gelatin micro and nano machine may be used to disperse pharmaceuticals due to polymer swelling. According to these findings, rolling micro and nanorobots can provide doxorubicin payloads when they are induced by near-infrared light. Consequently, they could detect target cancer cells using cell-specific antibodies, and swim against the blood circulation. Safety tests on 3D-printed gelation microrobots showed minimal cytotoxicity upon enzymatic disintegration. Drug release selectivity has also been increased by the direct implantation of pharmaceutical chemicals inside responsive materials to create controlled release. Layer by layer, for example, micro and nano rockets were built. The mixture had layers upon layers of chitosan and negatively and positively charged sodium alginate, which is where the medication doxorubicin was absorbed. The micro and nano rockets were positioned close to the HeLa cells and stuck to them to release the doxorubicin locally and cause apoptosis [13]. Vaccines are also known as a biological preparation that provide adequate immunity to a particular infectious disease and they have been a center of focus as target delivery using micro and nano machines. It has been found that living animals can get an effective vaccination using self-propelled nanobots, thus improving the efficacy of oral immunizations. The miniscule machines were about 20 μm long with polymer tubes coated with zinc and placed in the gut of a live mouse. Once zinc ions reacted with the acidic stomach, hydrogen generation propelled the nanobots to position onto the stomach lining. Besides, miniature machines can be powered by urea in addition to stomach acid. An intravenous infusion of the nanorobot was used to treat bladder cancer or infection. Magnesium-based micro machines discharged the vaccination to the target tissues. These microspheres were designed to be filled with three prominent layers, including chitosan, an enteric segment layer that responded to pH variations, and a red blood cell membrane that had an injected toxin [14].

8.5 Inorganic Material Agents

In addition to biological diminutive machine that use living things, such as DNA origami robots or protein motors to drive them, inorganic chemicals also exhibit a lot of promise in medicinal and therapeutic contexts. The use of living organism—motors in sensing or capturing purposes is constrained by their short lifetime and declining efficiency in complicated materials and complex fluid. Reasonably, complementary metallic structures are used for drug delivery and sensing in attempts to use synthetic inorganic materials as external triggers and responsive stimuli. Synergistic materials, such as metal organic framework (MOF), covalent organic framework (COF), and Mxenes with unique properties that can be adopted on-demand syntheses and utilized to minimize dose and therapeutic side effects without compromising therapeutic efficacy. Inherent magnetic properties of adhesive materials have been employed by magnetic-driven miniature machines to convert the kinetic energy into thermal energy by constantly applying external magnetic field. The heat also triggered the release of the anticancer drug 5-fluorouracil through temperature changes as well as the activation of the iron oxide clusters embedded in polymeric gel of micro transport machine [15]. The micro and nano machine as biosensors possess great promise in intelligent sensing and motion systems. Kong research team exploited a Mg/Pt Janus micromotor with self-invigorating surfaces for determining the variations of human serum glucose [16]. Furthermore, plaques in blood vessels of animals have recently been removed by using heat-activating micro machines. Previously, this intelligent machine achieved dual chemical and magnetic mobility by utilizing the well-defined structure, remarkable mechanical properties, and noteworthy electrical features of carbon nanotubes. Knifefish uses electricity to communicate through their electrocytes. The mobile microrobotic beads as given in Fig. 8.4a can be functionalized with the carboxyl-containing ligand poly(N-[3-(dimethylamino)propyl]methacrylamide) by EDC/NHS to achieve polymeric magnetic robots (see Fig. 8.4b). These magnetic robots abled to permanently trap bacteria after multiple washing in water as depicted in Fig. 8.4c [17].

A gelatine polyvinylidene fluoride (PVDF) was employed as a polymeric model to create a knifefish-shaped micro machine that can produce energy similarly with its electrocytes. The proposed model was further reconstructed Ni-doped polypyrrole to attain more multimodalities. Due to the electrostatic repulsion augmented by the piezoelectric impact, the surface of the constructed nano capsules provided a better drug discharge. Hence, the potential of piezoelectric performance of specialty plastic material, such as PVDF also manipulated the trigger and release rate of these capsules. The tiny bleeding wounds of living mice could be sealed and secured by

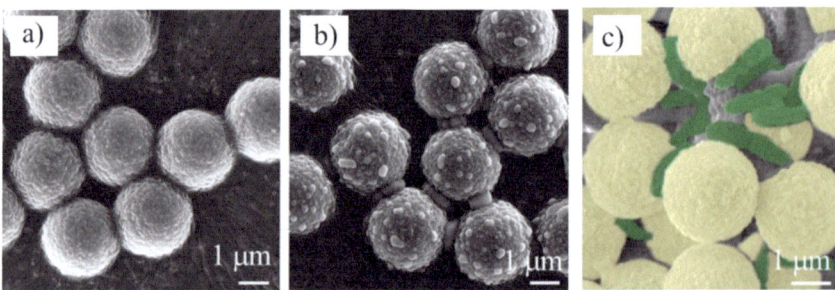

Fig. 8.4 Micro beads before (**a**) and after (**b**) the functionalization with the polymeric arms and **c** effectiveness of microrobots in trapping bacteria and microplastics in water. Reproduced with permission from Ref. [17]. Copyright 2024 American Chemical Society

8.5 Inorganic Material Agents

photothermal minute machines. Janus microstructure with spherical gold was capped with magnetite nanoparticles and fluorescent dyes as a minute robot. Subsequently, they were activated with infrared laser during thermophoresis process for tissue scaffold repair. More evidently, the miniature motor executed at localized target by applying external heat, which caused the endothelial matrices as biological adhesive to actively heal the wound and denature when a laser pulse was applied directly to the target [19]. Significant task is required to design and construct robots to meet the needs for clinical treatment. The multiscale magnetic hydrogel robot (MMHR) with a core–shell structure in which superparamagnetic nanoparticles and curcumin-integrated chitosan microparticles are condensed inside the core and doxorubicin-loaded chitosan microbeads are trapped inside the shell. The codelivery of doxorubicin and curcumin was realized through magnetic actuation system as depicted in Fig. 8.5a for oral administration applications. The magnetic driving system shown in Fig. 8.5b could manipulate the MMHR beads (see Fig. 8.5c) to achieve locomotive drive with high precision [18]. To advance a precision medicine, when coagulating agent could not be to trigger a series of proteolytic events that leads for fibrin formation at vascular damage despite heavy bleeding, the developed diminutive robot helped to heal the wound by minimizing coagulant concentration variations. Additionally, without the need of regular antibiotics, silver clusters can be transferred into the bacterial cytoplasm and rupture their outer membrane when capped on the minuscule machines to act as bactericidal agents. Because large-scale surgical equipment does not have comparable micro/nanoscale counterparts, it is challenging to undertake procedures at this small scale, which results in minimal tissue penetration. The miniaturization of surgical instruments may have noticeable advantages because of their small size and ability to fit into areas that blades and catheters cannot. To directly access or remove biological tissues, micro and nanorobotics may be employed as surgical tools. In fact, a micro and nanorobot might improve upon the features of the apparatus used in surgical robotics today, giving human surgeons more control and accuracy [20].

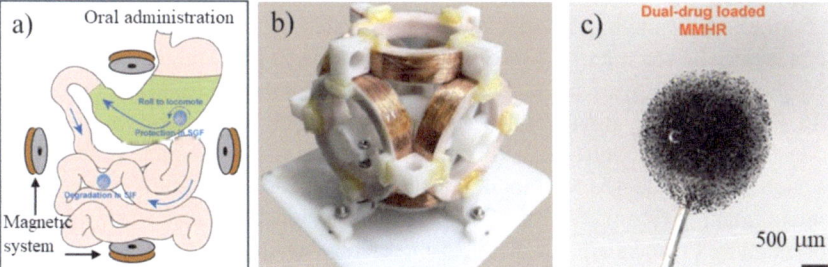

Fig. 8.5 **a** Schematic of magnetic actuation targeting system. **b** Triaxial Helmholtz coils. **c** Dual-drug loaded MMHR bead. Reproduced with a permission from Ref. [18]. Copyright 2022 American Chemical Society

8.6 Image Guidance for Monitoring Systems

Using an external light source to detect tissue reflection, optical imaging technology takes pictures. Among the different imaging procedures, it is the most straightforward and basic. Its features include ease of use, affordability, and quick imaging. However, its capacity to penetrate is restricted because of tissue absorption and light scattering. Because of its comparatively good tissue penetration capability, near-infrared fluorescence imaging technology has been widely used in vivo imaging to overcome the constraint of low penetration in fluorescence imaging. To in order to enhance the transfer of medical micro and nano machines from bench to clinical settings, imaging technologies contribute a significant role in enabling real-time tracking of a micro and nano machine, referred as MagRobots in vivo. Clinically proven imaging functionalities, including optical imaging, magnetic resonance imaging (MRI), near-infrared light (NIR) magnetic particle imaging (MPI), fluorescence imaging, ultrasound (US) imaging, photoacoustic imaging (PA), X-ray computed tomography (CT), photoacoustic computed tomography (PACT), optical coherence tomography (OCT), single photoemission computed tomography (SPECT), positron emission tomography (PET), and their combined imaging techniques, such as the integration of MR and CT, integration of PET and CT, and integration of PET and MRI, appear to be compatible with miniaturized robotics equipment. As a classical clinical imaging method, ultrasound imaging normally requires two different modes, including Doppler and B-mode. While the latter exploits the Doppler effect while the former is relied on the pulse-echo techniques. High temporal and spatial resolution, deep penetration, little tissue injury, and very cheap setup are imaging's primary benefits. As illustrated in Fig. 8.6a, near-infrared light (NIR)-responsive hollow magnetic nanocarriers (HMC) were integrated with a chitosan-based molecular valve onto hollow magnetic nanocarriers (CHMC) to enhance NIR-triggered drug loading. The mice injected with the CHMC nanoparticles and doxorubicin (DOX) shown in Fig. 8.6b revealed a drastic temperature increase in the presence of infrared light expose, whereas control group provided no significant temperature change [21].

Furthermore, imaging apparatus was incorporated to track and follow a swarm of magnetically driven micro machines inside a cow's eye [4]. The hairbots in the ex-vivo chicken breast were tracked using the color Doppler ultrasonography function [22]. Furthermore, endovascular delivery was engaged using the chosen Doppler mode for real-time control of a swarm of magnetic micro machines [23]. Gleich research team conducted MPI modalities to construct as a three-dimensional tomographic imaging technique [24]. Two permanent magnets positioned in a Maxwell reorientation to make up the MPI scanner. Relying on external field gradients, magnetic modules in the MPI scanning section can be propelled with obvious propulsion forces. However, due to the current platform's limited spatial resolution of MPI, which is just a few millimeters, this technique is only relevant to only visualization of swarming micro and nano machines. Tay and collaborators also conveyed the precise localization of magnetic hyperthermia, triggered by the integration of MPI

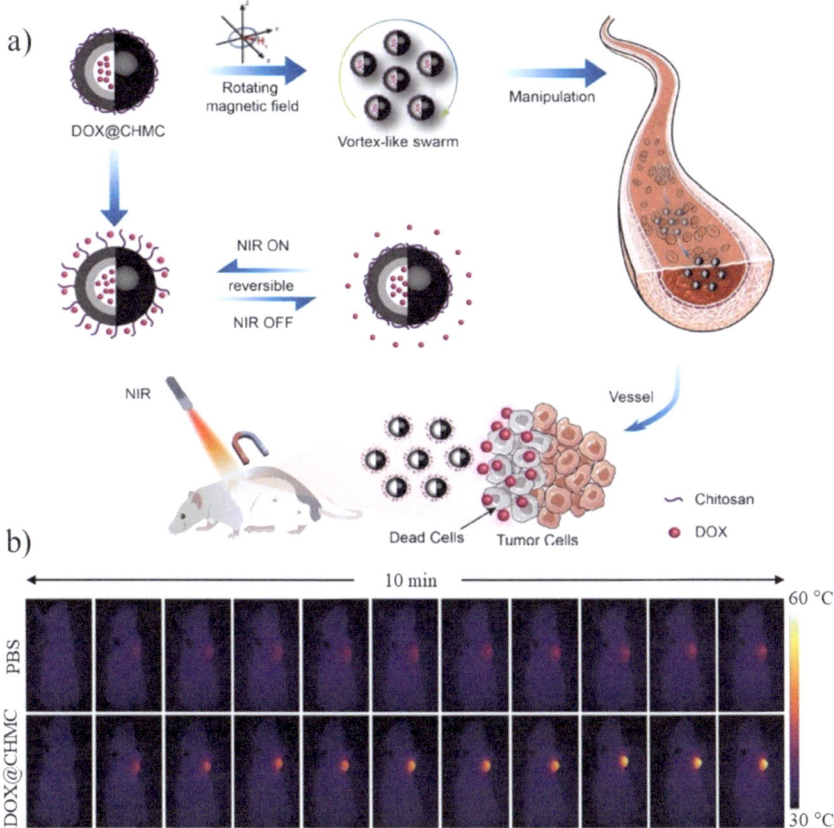

Fig. 8.6 a Representative diagram of magnetically driven CHMC micro and nanorobots swarms for targeted drug delivery, NIR-responsive drug loading and treatment of cancer. **b** Comparison of NIR thermal photographs of PBS and DOX@CHMC on mice under laser irradiation. Reproduced with a permission from Ref. [21]. Copyright 2024 American Chemical Society

gradient and superparamagnetic magnetic nanoparticles. It was also enhanced with quantitative navigation of MPI imaging to tumor sites of their selecting [25]. When the field-free region (FFR) of the MPI gradient was directed to the targeted tumor location, the immediate heat only neutralized the malignant tissues while minimizing the collateral heat damage to neighboring healthy tissues. Superparamagnetic nano machines situated in the liver and tumor of a U87MG xenograft mouse were mapped to experimentally illustrate an image-tracked theragnostic model using a combination of magnetic particle imaging and localized magnetic hyperthermia [4].

8.7 Untethered Micro and Nanomachines for Remote Intelligent Sensing

Industrial development and cumulative anthropogenic activities cause unsafe chemicals to be released into the natural environment. These complexes can be particularly toxic, can deteriorate human bodies, and harm the marine environment. Therefore, methods for sensing and determining these toxins and pollutants, such as viruses, heavy metal compounds, and bacterial endotoxins, becoming important for food security and health monitoring. Although sensing systems trace variations in the target analyte under diverse conditions, detection methods provide information on the presence and concentration of the target analyte. Conventional methodologies usually execute either indirect determination with probe techniques (e.g., aptamer probes) or direct evaluation of toxins (e.g., high-performance liquid chromatography (HPLC)). These procedures normally demand sampling, thorough specimen decontamination, and skilled operation of advanced equipment. In the incident of difficult-to-reach target and easily polluted samples, in situ sensing using diminutive machines is beneficial, particularly for real-time monitoring of drastically varying concentrations of target analytes or toxins. In addition, using established sensing approaches meets active operation-based sensing difficult to measure, such as assessing the local mechanical section of biological cells and tissues. With the advance in this sensing, nanorobot-generated signals, such as on–off fluorescence for toxin recognition, on–off luminescence for chemical sensing, and electro chemiluminescent for biosensing, are the footing of self-generated signal-based sensing. For instance, the unique surface features of the micro carriers or machines have a significant influence on their sensing performance. When compared to the MoS_2 and Pt-integrated micro machines, the WS_2 and Pt-integrated micro machines with a thicker outer surface illustrates almost 4 folds increase in sensitivity because of more effective peptide probe payload and release. Consequently, with a detection limit of 1.9 ng mL^{-1}, the peptide-modified WS_2 and Pt-integrated micro machines are exploited as an inexpensive sensing tool for the high-throughput measurement of *Escherichia coli* endotoxin. A capping fluorophore fluoresceinamine (FLA)/silica and NH_2/Pt-integrated micro machine is investigated for on-the-hover sensing of sarin and soman simulants. It was also based on the on–off fluorescent techniques, in addition to the cylindrical-shaped micro machine-based sensing method [27].

Fluorescent magnetic spore-based diminutive carriers have been developed as mobile sensors for detecting toxins released by C. diff in patients' stool in addition to the self-driven micro machine-based sensing approach [28]. For C-reactive protein (CRP) detection, rGO-based magnetic diminutive machines are bound to an electrochemical microfluidic chip, as given in electrochemical microfluidic setup in Fig. 8.7a. The diminutive machine illustrates tubular shape with membrane porous structure as given in Fig. 8.7b. Sandwich immune-complexation was conducted by the antibody-grated microrobots using CRP and horseradish peroxidase (HRP)-labeled anti-CRP secondary antibody in an outer reservoir shown in Fig. 8.7c. The mechanism of electrochemical detection later facilitated by their subsequent magnetic retaining

8.7 Untethered Micro and Nanomachines for Remote Intelligent Sensing

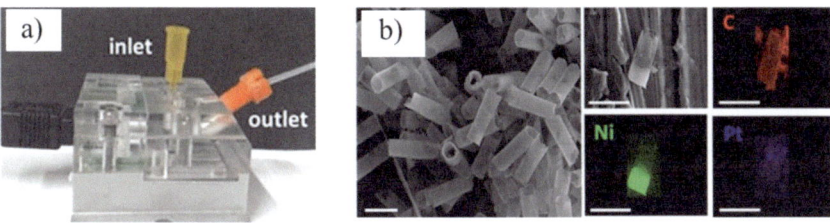

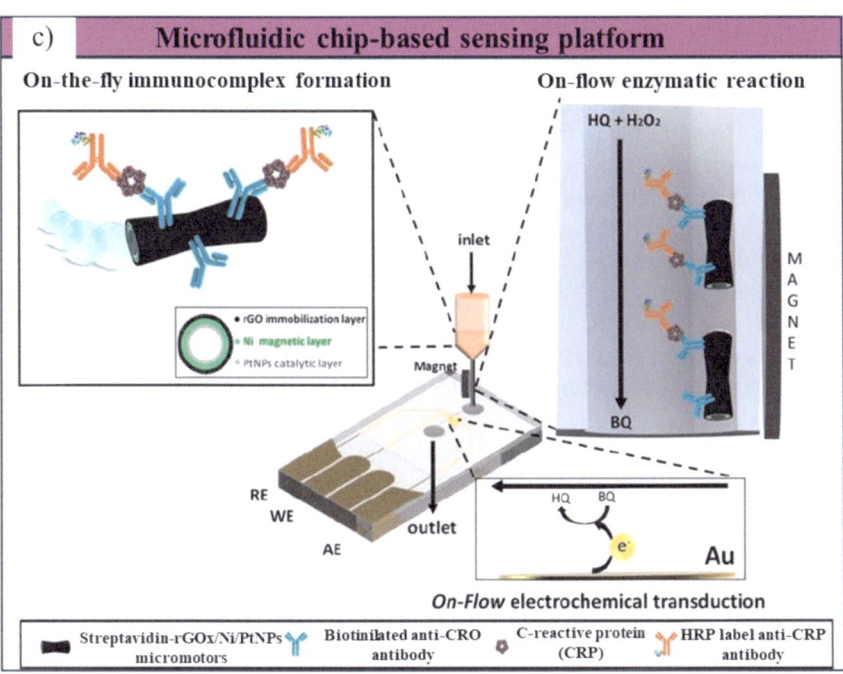

Fig. 8.7 **a** Electrochemical microfluidic setup with external reservoir (inlet) and thread connection (outlet). **b** SEM image and EDX analysis of rGO/Ni/PtNPs micromotors with a scale bar of 10 μm. **c** For CRP sensing, electrochemical microfluidic chips use magnetic reduced graphene oxide (rGO)-based micro machines. Adapted with a permission from Ref. [26]. Copyright 2020 American Chemical Society

in the metallic collector surface during the passage of enzyme on substrates. Automated, sensitive, and selective CRP detection without decreasing performance was made possible by the combination of the micro carrier system and microfluidic electrochemical detection technique [26]. Micro and nano machine advancement were also amalgamated into mobile systems with miniaturized chips to create real-time measurement, portable, and affordable disease testing apparatus [29]. Currently, most of the sensing and detection with micro nanorobots is carried out in vitro. Nevertheless, the sophisticated operating surroundings presents it hard to indeed identify signals and track tiny machines' behavior. Hence, it makes it difficult to manipulate and image these machines inside a living organism. In addition, more research is needed to adopt methods for the setting deployment and recycling of the machines. Using micro and nano machines for in vivo monitoring permits for the real-time determination of in vivo data through machine-produced signals and performances, markedly in the complicated surroundings of living organisms. This research field also carry out the benefits of their miniature size and controllable transportation. It assists with early illness prognosis by enabling us to gather information about target locations without needing invasive laborious procedures. Nonetheless, promising biocompatibility issues give it challenging to implement chemical fuel-propelled micro nano carriers for in vivo applications. By exploiting other possible propulsion strategies, such as magnetic or ultrasonic trigger and using biocompatible compounds to fabricate tiny machines, these concerns can be resolved.

In addition, real-time in vivo tracking of micro and nano machines relies on obtaining high imaging contrasts, which describes for the integration of several medical imaging functions to visualize the mobile micro and nano machines, acquiring data for sensing. Since micro nano machines are vital when integrating with the best actuation type and imaging modality, their design and functionalization ought to be given top importance. An approach for the signal processing and sensing mechanism should be evaluated and depended on environmental variations and sensing tasks. This strategy also guides the design of the tiny machine system. To achieve an effective and selective system with a certain degree of automate detection, a trade-off between target and interferent elements should be considered from the standpoint of system integration. Most micro and nano transporting machines on the market today are made to spot a single object. Convenient and fast identification of several targets will be made possible by the advancement of multifaceted sensing platforms. Future work will focus on advancing intelligent sensing techniques through artificial intelligent-based platforms. It can overcome the mentioned obstacles, which will need an integration between several artificial intelligence disciplines and relevant technological advancements in micro and nano machines. Since propulsion of micro and nano machines are central when combining with external actuation type and imaging techniques. Their design and functionalization with suitable chemical should be consider as a primacy choice.

References

1. C. Hu, S. Pané, B.J. Nelson, Soft micro- and nanorobotics. Annu. Rev. Control Robot. Auton. Syst. **1**(1), 53–75 (2018)
2. B.J. Nelson, L. Dong, F. Arai, Micro-/nanorobots, in *Springer Handbook of Robotics* (2016), pp. 671–716
3. B. Wang, Y. Zhang, L. Zhang, Recent progress on micro- and nano-robots: towards in vivo tracking and localization. Quant. Imaging Med. Surg. **8**(5), 461 (2018)
4. H. Zhou et al., Magnetically driven micro and nanorobots. Chem. Rev. **121**(8), 4999–5041 (2021)
5. A.V. Chesnitskiy et al., Bio-inspired micro- and nanorobotics driven by magnetic field. Materials **15**(21), 7781 (2022)

References

6. T. He, Y. Yang, X.-B. Chen, Propulsion mechanisms of micro/nanorobots: a review. Nanoscale (2024)
7. K. Chen et al., A Pt/Au hybrid self-actuating nanorobot towards to drug delivery system, in *10th IEEE International Conference on Nano/Micro Engineered and Molecular Systems* (IEEE, 2015)
8. H. Zhang et al., Review of the applications of micro/nanorobots in biomedicine. ACS Appl. Nano Mater. **7**(15), 17151–17192 (2024)
9. A. Manjunath, V. Kishore, The promising future in medicine: nanorobots. Biomed. Sci. Eng. **2**(2), 42–47 (2014)
10. X. Sun et al., A soft capsule for magnetically driven drug delivery based on a hard-magnetic elastomer foam. ACS Biomater. Sci. Eng. **9**(12), 6915–6925 (2023)
11. M. Suhail et al., Micro and nanorobot-based drug delivery: an overview. J. Drug Target. **30**(4), 349–358 (2022)
12. P. Mena-Giraldo, J. Orozco, Polymeric micro/nanocarriers and motors for cargo transport and phototriggered delivery. Polymers **13**(22), 3920 (2021)
13. W. Chen et al., Recent progress of micro/nanorobots for cell delivery and manipulation. Adv. Funct. Mater. **32**(18), 2110625 (2022)
14. D.S. Grewal, Nanorobots as an alternative for treatment to vaccines for COVID-19. Historian **1** (2010)
15. F. Soto et al., Medical micro/nanorobots in precision medicine. Adv. Sci. **7**(21), 2002203 (2020)
16. L. Kong et al., Micromotor-assisted human serum glucose biosensing. Anal. Chem. **91**(9), 5660–5666 (2019)
17. M. Ussia et al., Magnetic microrobot swarms with polymeric hands catching bacteria and microplastics in water. ACS Nano **18**(20), 13171–13183 (2024)
18. X. Chen et al., Multiscale magnetic hydrogel robot with a core–shell structure for active targeted delivery. ACS Appl. Polym. Mater. **4**(11), 8645–8655 (2022)
19. M. Koleoso et al., Micro/nanoscale magnetic robots for biomedical applications. Mater. Today Bio **8**, 100085 (2020)
20. M. Urso, M. Pumera, Micro- and nanorobots meet DNA. Adv. Funct. Mater. **32**(37), 2200711 (2022)
21. X. Chen et al., Hollow magnetic nanocarrier-based microrobot swarms for NIR-responsive targeted drug delivery and synergistic therapy. ACS Appl. Mater. Interfaces **16**(44), 60874–60883 (2024)
22. A.V. Singh et al., Multifunctional magnetic hairbot for untethered osteogenesis, ultrasound contrast imaging and drug delivery. Biomaterials **219**, 119394 (2019)
23. Q. Wang et al., Ultrasound Doppler-guided real-time navigation of a magnetic microswarm for active endovascular delivery. Sci. Adv. **7**(9), eabe5914 (2021)
24. B. Gleich, J. Weizenecker, Tomographic imaging using the nonlinear response of magnetic particles. Nature **435**(7046), 1214–1217 (2005)
25. Z.W. Tay et al., Magnetic particle imaging-guided heating in vivo using gradient fields for arbitrary localization of magnetic hyperthermia therapy. ACS Nano **12**(4), 3699–3713 (2018)
26. A.G. Molinero-Fernández, M.A.N. López, A. Escarpa, Electrochemical microfluidic micromotors-based immunoassay for C-reactive protein determination in preterm neonatal samples with sepsis suspicion. Anal. Chem. **92**(7), 5048–5054 (2020)
27. V. de la Asunción-Nadal et al., Chalcogenides-based tubular micromotors in fluorescent assays. Anal. Chem. **92**(13), 9188–9193 (2020)
28. Y. Zhang et al., Real-time tracking of fluorescent magnetic spore–based microrobots for remote detection of C. diff toxins. Sci. Adv. **5**(1), eaau9650 (2019)
29. M.S. Draz et al., DNA engineered micromotors powered by metal nanoparticles for motion based cellphone diagnostics. Nat. Commun. **9**(1), 4282 (2018)

MIX
Papier aus verantwortungsvollen Quellen
Paper from responsible sources
FSC® C105338

If you have any concerns about our products,
you can contact us on
ProductSafety@springernature.com

In case Publisher is established outside the EU,
the EU authorized representative is:
**Springer Nature Customer Service Center GmbH
Europaplatz 3, 69115 Heidelberg, Germany**

Printed by Libri Plureos GmbH
in Hamburg, Germany